AF683003

Lecture Notes in Intelligent Transportation and Infrastructure

Series Editor

Janusz Kacprzyk, Systems Research Institute, Polish Academy of Sciences, Warsaw, Poland

The series “Lecture Notes in Intelligent Transportation and Infrastructure” (LNITI) publishes new developments and advances in the various areas of intelligent transportation and infrastructure. The intent is to cover the theory, applications, and perspectives on the state-of-the-art and future developments relevant to topics such as intelligent transportation systems, smart mobility, urban logistics, smart grids, critical infrastructure, smart architecture, smart citizens, intelligent governance, smart architecture and construction design, as well as green and sustainable urban structures. The series contains monographs, conference proceedings, edited volumes, lecture notes and textbooks. Of particular value to both the contributors and the readership are the short publication timeframe and the world-wide distribution, which enable wide and rapid dissemination of high-quality research output.

More information about this series at http://www.springer.com/series/15991

Anna Kwasiborska · Jacek Skorupski ·
Irina Yatskiv
Editors

Advances in Air Traffic Engineering

Selected Papers from 6th International Scientific Conference on Air Traffic Engineering, ATE 2020, October 2020, Warsaw, Poland

Springer

Editors
Anna Kwasiborska
Faculty of Transport, Division of Air
Transport Engineering
Warsaw University of Technology
Warsaw, Poland

Jacek Skorupski
Faculty of Transport, Division of Air
Transport Engineering
Warsaw University of Technology
Warsaw, Poland

Irina Yatskiv
Transport and Telecommunication Institute
Riga, Latvia

ISSN 2523-3440 ISSN 2523-3459 (electronic)
Lecture Notes in Intelligent Transportation and Infrastructure
ISBN 978-3-030-70923-5 ISBN 978-3-030-70924-2 (eBook)
https://doi.org/10.1007/978-3-030-70924-2

© The Editor(s) (if applicable) and The Author(s), under exclusive license
to Springer Nature Switzerland AG 2021
This work is subject to copyright. All rights are solely and exclusively licensed by the Publisher, whether the whole or part of the material is concerned, specifically the rights of translation, reprinting, reuse of illustrations, recitation, broadcasting, reproduction on microfilms or in any other physical way, and transmission or information storage and retrieval, electronic adaptation, computer software, or by similar or dissimilar methodology now known or hereafter developed.
The use of general descriptive names, registered names, trademarks, service marks, etc. in this publication does not imply, even in the absence of a specific statement, that such names are exempt from the relevant protective laws and regulations and therefore free for general use.
The publisher, the authors and the editors are safe to assume that the advice and information in this book are believed to be true and accurate at the date of publication. Neither the publisher nor the authors or the editors give a warranty, expressed or implied, with respect to the material contained herein or for any errors or omissions that may have been made. The publisher remains neutral with regard to jurisdictional claims in published maps and institutional affiliations.

This Springer imprint is published by the registered company Springer Nature Switzerland AG
The registered company address is: Gewerbestrasse 11, 6330 Cham, Switzerland

Preface

The 6th International Conference on Air Traffic Engineering (ATE) was held during 20 October 2020. The ATE conference first took place in 2010 and was intended as a forum for discussion of contemporary problems and scientific solutions in air traffic control. This tradition continues to date.

The conference focused on air traffic management, airports planning and management, safety, reliability and risk analysis, new air navigation technologies, legal and economic aspects of air transport, the human factor in air traffic engineering and also the environmental impact of air transport.

The conference was organized by the Department of Air Transport Engineering and the Students' Scientific Association of Air Transport (SKNTL) of the Warsaw University of Technology. It was held under the patronage of the IATA Association, the Civil Aviation Authority, the Ministry of Infrastructure and the Rector of the Warsaw University of Technology. The leading industrial partner of the conference was the Solidarity Transport Hub Poland, as well as ARUP Poland, LOT Polish Airlines, Polish Air Navigation Services Agency and LS Airport Services.

The conference was initially supposed to be held in the traditional formula. Unfortunately, the limitations of the COVID-19 pandemic caused it ultimately being organized entirely through electronic media. The event included representatives from academia and industry as well as practitioners. It was divided into two parts. The first part concerned innovative solutions in the field of air transport and relations with other modes of transportation, as well as the impact of the planned construction of the Solidarity Transport Hub Poland on the labour market. The second part was strictly scientific and included presentations of papers prepared by scientists working on the problems covered by the topics of the conference.

Texts of articles submitted for the conference were accepted for publication in the conference proceedings based on at least two independent positive reviews. The review took place under the "single-blind" procedure, so the authors did not know the names of people assessing their work. Out of the 35 papers submitted, only 12 were accepted for publication in LNITI.

As chair of the scientific committee and on behalf of members of both scientific and organizational committees, I want to thank all the collaborators—invited speakers, papers' presenters, reviewers of manuscripts and last but not least participants of the conference. They have all contributed to the success of this meeting.

October 2020 Jacek Skorupski

Organization

Organizers

Division of Air Transport Engineering, Faculty of Transport, Warsaw University of Technology
Student Research Group of Air Transport

Scientific Program Committee

Jacek Skorupski	Warsaw University of Technology, Poland
Teresa Abramowicz-Gerigk	Gdynia Maritime University, Poland
Martin Bugaj	University of Zilina, Slovakia
Alexei Sharpanskykh	Delft University of Technology, The Netherlands
Irina Jackiva	Transport and Telecommunication Institute, Latvia
Andrzej Fellner	Silesian University of Technology, Poland
Iwona Grabarek	Warsaw University of Technology, Poland
Jerzy Manerowski	Warsaw University of Technology, Poland
Sławomir Michalak	Air Force Institute of Technology, Poland
Luboš Janko	Czech Technical University in Prague, The Czech Republic
Michelle Bandeira	Instituto Tecnologico de Aeronautica, Brazil
Andrej Novak	University of Zilina, Slovakia
Alena Novak-Sedlackova	University of Zilina, Slovakia
Piotr Uchroński	WSB University, Poland
Mirosław Siergiejczyk	Warsaw University of Technology, Poland
Anna Stelmach	Warsaw University of Technology, Poland
Olegas Prentkovskis	Vilnius Gediminas Technical University, Lithuania
Krzysztof Zboiński	Warsaw University of Technology, Poland

Contents

Mathematical Model of the Vehicle Exploitation System of the Tactical Air Base

Tomasz Chukowski[1], Jarosław Bartoszewicz[2](✉), and Marcin Kiciński[2]

[1] 1st Tactical Air Wing in Swidwin, Polczynska 32 Street, 78-301 Swidwin, Poland
[2] Faculty of Environmental and Energy Engineering, Poznan University of Technology, Piotrowo Street 3, 60-965 Poznan, Poland
{jaroslaw.bartoszwicz,marcin.kicinski}@put.poznan.pl

Abstract. This paper discusses vehicle problems in Tactical Air Base (TAB). The authors characterized the vehicle exploitation system of the polish armed forces and TAB. The aim of the authors was to propose a mathematical model of specific characteristics of BLT motor vehicle exploitation system. They present some indicators which should enable the assessment of the exploitation system. Selected values of the operational indicators determined by the test method are presented. Statistical research enabling understanding of the functioning dependencies in the examined exploitation system was introduced. The impact of the technical material supply system on the TAB vehicle exploitation system was examined. In turn, the authors present the mathematical model of the TAB vehicle exploitation system and the stages of its development. Next, calculation experiments using the created mathematical model were discussed. Obtained results can be used to support decisions in the field of optimizing the vehicle exploitation system. Striving to improve the operation system, i.e. improving certain indicators, involves making certain decisions.

Keywords: Tactical Air Base · Vehicle exploitation system · Mathematical model

1 Research Problem

Polish Armed Forces (PAF) is undergoing modernization in order to adapt them to current requirements. In the area of the Military Logistics System (MLS), the territorial support system of the PAF is used. Its basic function is to use the regional logistics base (RLB) and garrison support units (GSU), whose task is to provide logistical and financial support for military units [6, 15–18]. Tactical air bases (TAB) extend the scope of tasks and reorganization of structures. Currently, TAB, as one of the operational units, performs operational and logistics support tasks, such as GSU. From the perspective of several years of the new regulations being in force, it can be assessed that TAB has met the challenges and successfully implements logistics support tasks in the designated area of responsibility. The modernization of the MLS and further changes should continue to

© The Author(s), under exclusive license to Springer Nature Switzerland AG 2021
A. Kwasiborska et al. (Eds.): ATE 2020, LNITI, pp. 1–10, 2021.
https://doi.org/10.1007/978-3-030-70924-2_1

increase the effectiveness of the logistic support of the armed forces. One of the elements of the functional logistics system of the Armed Forces is the technical subsystem [4, 6]. Management of the use of military equipment is a very important function of the technical subsystem that requires further modernization. Modern IT tools can be very helpful in effective management of the operation of military equipment [5]. The purpose of organization and processing of operations is to create such organizational and technical conditions in the PAF that military equipment (ME) can be used at the planned time, place and with a certain intensity. The main operational management criteria [2, 10–12] include: technical readiness of ME (including technical performance), safety, economy of operation. Operating scenarios of the ME used in the Polish Armed Forces include: an exploitation strategy consistent with consumption over the period of use, an operation strategy consistent with the actual state. According to the instruction DD 4.22 (A) [7, 8, 14], the head of the military unit responsible for logistic support of the Polish Armed Forces manages the operation phase of the ME in the PAF. The abovementioned military units are responsible for the support and assurance stage, as well as defining the principles and standards for the operation of the ME, in cooperation with a competent ME administrator, as well as supervising compliance with operating rules. Depending on the scope of operations and period of use, equipment maintenance is divided into the following types [9, 11, 12, 19]: current service (pre-departure inspection, inspection in progress, service after use) and periodic maintenance (after mileage or storage time): periodic maintenance No. 1, periodic maintenance No. 2, subsequent periodic maintenance, special handling).

The scope of service depends on the type of service, type of ME and its current technical condition. The scope of service activities is specified in the technical documentation. ME repair involves organizational and technical activities aimed at restoring the utility function of ME by removing failures and damages resulting from use, or restoring life. Depending on the place of implementation, repairs are carried out by: Users (ME users), logistic units (such as GSU, RLB, air bases) and the national potential of industrial defense, including the program of economic mobilization and the potential of the allied foreign industry. Due to the scope, the following types of repairs are performed [1, 2, 9, 11, 13, 19]: current repair (CR), medium repair (MeR), main repair (MaR), maintenance repair (MtR), docking station repair (DR), emergency repair (ER), warranty repair (WR), consequential repair (RR) and combat damage repair (R1… R5). From the qualification point of view, repairs are divided into planned and unplanned. Planned repairs are MeR, MaR, MtR and DR. The scope of planned repairs is based on ME technical documentation. Unplanned repairs are CR, ER, WR and RR. Service standards and working time are specified in the catalogs of ME operating standards, ME technical documentation and technological guides, etc. In the Polish Armed Forces, the district supply rule applies. The supplying unit is GSU at the air base, which is done in accordance with the economic allocation schedule of the Ministry of National Defense. As part of the supply of technical materials, planned and unplanned deliveries are carried out. Planned materials are delivered hierarchically, i.e. the higher the level, the lower the level, through the GSU of the air base supplying the military unit and institutions within their responsibility, through the RLB supplying the GSU based on the approved annual delivery plan and unplanned supplies (military unit and institutions submit applications to GSU, and then technical materials are delivered from warehouses. In the absence of deliveries, the purchase is carried out by a subordinate unit).

2 Methods

The article will present indicators. This group of indicators is enough to assess the vehicle exploitation system and includes, among others: the technical availability indicator Kg [–], the vehicle parking indicator for maintenance and repairs Kp [–], the burden factor at service station N [–], the vehicle standstill time for servicing and repairs per 1000 km Ton' [h/1000 km], the average number of days of the vehicle being in servicing and repairs Ton_poj_sr [days], the real average service and repair time Ton_rzecz_sr [h], the average passive downtime in service and repair Ton_pos_sr [h], the average service and repair duration Ton_sr [h], the passive downtime indicator in service and repair Kpos [–]. The technical availability indicator Kg is the basic indicator of the reliability of a technical object involved in the operation process. In the case of mobile assets consisting of motor vehicles, the technical readiness indicator is determined according to the following formula:

$$\mathrm{Kg}(t) = \frac{\sum_i T_i^{\mathrm{u}}(t)}{\sum_i T_i^{\mathrm{u}}(t) + \sum_i T_i^{\mathrm{on}}(t)} [-], \tag{1}$$

where: T_i^{on} is the total time of failure of the i-th vehicle, and T_i^{u} is the total time of the i-th vehicle use. Another important indicator from the vehicle exploitation system point of view is the so-called vehicle parking indicator for maintenance and repairs Kp [–], which is determined from the formula:

$$\mathrm{Kp}(t) = \frac{\sum_i T_i^{\mathrm{on}}(t)}{\sum_i T_i^{\mathrm{u}}(t) + \sum_i T_i^{\mathrm{on}}(t)} [-]. \tag{2}$$

In the situation of technical facilities, the burden factor at service station N plays a special role. It can be determined from the formula:

$$N = \frac{\sum_i T_i^{\mathrm{on\_dr}}}{T^{\mathrm{hor\_wl}}} [-], \tag{3}$$

where: Thor_wl is the total availability of service stations [h] and Ton_dr is the total duration of service and repair at the service station [h]. Having the mileage of the vehicle, the average vehicle standstill time for servicing and repairs per 1000 km Ton' can be determined according to the below formula:

$$T^{on'}(t) = 1000 \cdot \frac{T^{on_{d}r}(t)}{l} \left[\frac{h}{1000\,km}\right]. \tag{4}$$

where: l is the vehicle mileage for a period of time [km].

The average number of (working) days of the vehicle being in repairs and servicing $T^{\text{on\_poj\_sr}}$ is calculated from the formula:

$$T^{\text{on\_poj\_sr}}(t) = \frac{T^{\text{on\_dr\_sr}}(t)}{7 \cdot s}[\text{days}]. \tag{5}$$

where s is number of vehicles [–].

Considering the service or repair stages, it is worth using time indicators such as:

- real average service and repair time $T^{\text{on\_rzecz\_sr}}$ [h] determined from the formula:

$$T^{\text{on\_rzecz\_sr}}(t) = \frac{T^{\text{on\_rzecz\_sr}}(t)}{n}[\text{h}]. \tag{6}$$

where: $T^{\text{on\_rzecz}}(t)$ is the total time of real service and repair in the period of time and.
n is the number of services and repairs in the period of time,

- average passive downtime in service and repair $T^{\text{on\_pos\_sr}}$ [h] described by the equation:

$$T^{\text{on\_pos\_sr}}(t) = \frac{T^{\text{on\_pos}}(t)}{n}[\text{h}]. \tag{7}$$

Average service and repair duration Ton_sr [h] is the sum of the real average service and repair time Ton_rzecz_sr [h] and the average passive downtime in service and repair Ton_pos_sr [h]

$$T^{\text{on\_sr}} = T^{\text{on\_rzecz\_sr}} + T^{\text{on\_pos\_sr}} \tag{8}$$

From the point of view of the efficiency of the operation system, an important characteristic of the technical facilities is the passive downtime indicator in repair and service K^{pos} [–]:

$$K^{\text{pos}} = \frac{\sum_i T^{\text{on\_pos}}(t)}{\sum_i T_i^{\text{on\_dr}}(t)}[-], \tag{9}$$

As part of the research, all service and repairs recorded in the Technical Service Cards in 2017 which related to vehicles belonging to the TAB were analyzed. The data included in TCS allowed for calculation of the indicators characterizing the vehicle exploitation system described in article. The results of the operation indicators for subunits and the TAB as a whole are listed in Table 1. In order to further investigate the problem, service and repair were identified as two separate processes.

Duration of service and repair is the difference between the dates of service and repair start and end recorded in Technical Service Cards (TSC). The above time is presented in hours, assuming that one working day includes 7 h of work at the service station. The actual time of repair or service results from the man hours and number of employees carrying out repair or service recorded in TSC and reflects the actual time of work at the service station. Passive downtime in service and repair results from the difference between the duration of service and repair and the actual time of repair and service. After comparing the data in the Table 1, it can be clearly seen that the times and labor consumption of service differ significantly from the times and labor consumption of repair, which seems to be understandable, considering the different scopes of operation. Another regularity, which clearly comes from the presented results, is the extreme difference in passive downtime between service and repair.

Table 1. Operation indicators

Name of indicator	1st PTC*	2nd PTC	3rd PTC	AGHP	ASC	SV	TAB
Milage 2017 [thousand km]	486.7	196.2	18.2	34.7	76.7	63.0	875.6
Average consumption							
– service $T^{o\_rbh\_sr}$ [man hours]	10.14	11.93	10.50	10.00	8.27	12.44	10.13
– repair $T^{n\_rbh\_sr}$ [man hours]	24.41	30.27	19.25	28.50	35.60	18.00	24.56
Real average time of							
– service $T^{o\_rzecz\_sr}$ [h]	8.93	9.68	7.00	9.50	7.32	10.89	8.62
– repair $T^{n\_rzecz\_sr}$ [h]	20.69	23.42	14.82	24.03	30.55	15.57	20.80
Average service duration T^{o_sr} [h]	10.40	14.50	7.00	13.00	10.50	14.78	11.47
Average repair duration T^{n_sr} [h]	51.07	51.42	31.50	65.06	80.85	35.20	48.06
Average passive downtime in							
– service $T^{o\_pos\_sr}$ [h]	1.47	4.82	0.00	3.50	3.18	3.89	2.85
– repair $T^{n\_pos\_sr}$ [h]	30.39	28.00	16.68	41.03	50.30	19.63	27.33
Passive downtime indicator in							
– service K^{pos_o} [–]	0.14	0.33	0.00	0.27	0.30	0.26	0.25
– repair K^{pos_n} [–]	0.59	0.54	0.52	0.63	0.62	0.56	0.57
Average number of days the vehicle in service and repair $T^{on\_poj\_sr}$ [days]	16.70	9.58	6.40	11.48	9.42	6.96	9.46
Vehicle standstill time for servicing and repairs per 1000 km $T^{on'}$ [h/1000 km]	7.21	8.88	49.16	83.82	24.10	21.66	12.94
Vehicle parking indicator for maintenance and repairs K^{pos} [–]	0.059	0.034	0.022	0.031	0.035	0.025	0.034
Technical availability indicator Kg [–]	0.94	0.97	0.98	0.97	0.96	0.98	0.97

PTC – a platoon of the transport company, AGHP – the Aircraft Ground Handling Platoon, ASC – Airfield Service Company, SV – special vehicles, TAB – Tactical Air Bases

The results presented in Table 1 indicate a higher average number of days a vehicle was in service and repair Ton_poj_sr in subunits with higher annual mileage. This, in turn, results in the technical availability indicator Kg for these subunits being the lowest according to the formula (1) described in the article. The technical availability indicator Kg mentioned before describes the so-called stationary technical readiness independent of vehicle mileage. Considering the above, highly damaged vehicles can also show a high degree of stationary technical readiness with low intensity of use. To get a full picture related to vehicle reliability, one should also pay attention to vehicle standstill time for servicing and repairs per 1000 km Ton'. Analyzing the Ton' indicator, it can be seen that it reaches the highest values for 3rd PTC of the transport company and AGHP with low mileage.

3 Results

In order to determine the relationships in the operating system, many studies were conducted, which resulted in monthly time charts of various types of operational indicators showing their values in individual months within one year. One example graph is shown in Fig. 1a.

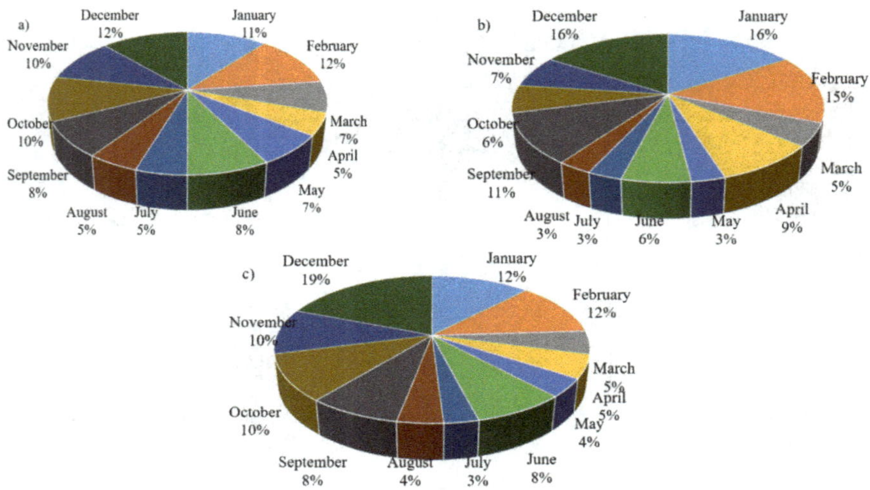

Fig. 1. A Time characteristics: a – repair time [days], b – average waiting time for spare parts [h], c – average monthly passive downtime in repair [h]

The impact of actual service and repair work changes as well as passive service and repair downtime on changes in total service and repair time was examined. The analysis of monthly time charts and the calculated correlation indicators show that the change in the average passive downtime in repair Tn_pos_sr has a significant impact on the changes in the total repair time Tn (the correlation coefficient is 0.9), in the case of the average real repair time Tn_rzecz_sr and the total repair time Tn, the correlation coefficient is only 0.39, between the passive downtime in the To_pos service and the total time of the service, it has no high correlations and the changes in the actual time of the To_rzecz service have a decisive impact on the changes in the total time of the service To (the correlation coefficient is 0.96). Given the very different opinions associated with the supply of spare parts, it was decided to investigate the impact of the technical material supply system on the vehicle exploitation system in TAB. After comparing the monthly average waiting time for spare parts (Fig. 1b) with the chart of the average monthly passive downtime in repair Tn_pos_sr created on the basis of archived TSC (Fig. 1c), it turned out that the above charts show a high correlation of 0.82. Analyzing the results of the average annual waiting time for spare parts (24.3 h) and the average annual passive downtime in repair (27.3 h), and their monthly distribution, it can be stated that the time associated with the supply of technical materials has a significant impact on the time of passive downtime in repair, and thus on the functioning of the TAB vehicle exploitation system.

The mathematical model (MM) of the vehicle exploitation system was created based on the experimental method widely described in the literature and on the basis of experience gained in the course of research. The diagram of the mathematical model creating process is shown in Fig. 2.

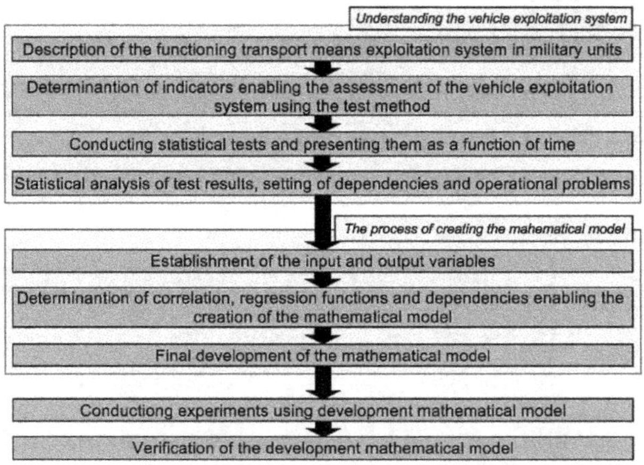

Fig. 2. The diagram of the mathematical model (MM) creating process

The MM created for the TAB vehicle exploitation system using the linear regression function takes into account input variables, which can be modified in the decision-making process. When choosing input variables, the operational problems have been taken into account. This problems can be reduced using proper input. On the other hand, output variables are indicators which allow to evaluate the functioning of the vehicle exploitation system when the input variables are modified.

The mathematical model presented in Fig. 3 was developed using formulas described in the article and others widely known in literature [2, 9, 11, 19]. The aim of the experiment was to examine the output quantities values (operation indicators) with changes of the input quantities values (factor of employee elevation gain at the service station P and the average waiting time for delivery of spare parts Tzap_sr).

Conditional analysis of data was used in the experiment. The modelled values of each exploitation indicator (output variable) are presented in the form of a data table with two input variables. The modeled values of average waiting time for delivery of spare parts T^{zap_sr} is presented as rows of cells and the modeled values of employee elevation gain at the service station P is presented as columns of cells (Fig. 4).

The results of the experiments were also depicted in the form of three-dimensional charts (Fig. 5). Due to the limited size of the article, only modelled values of the technical readiness indicator Kg are presented.

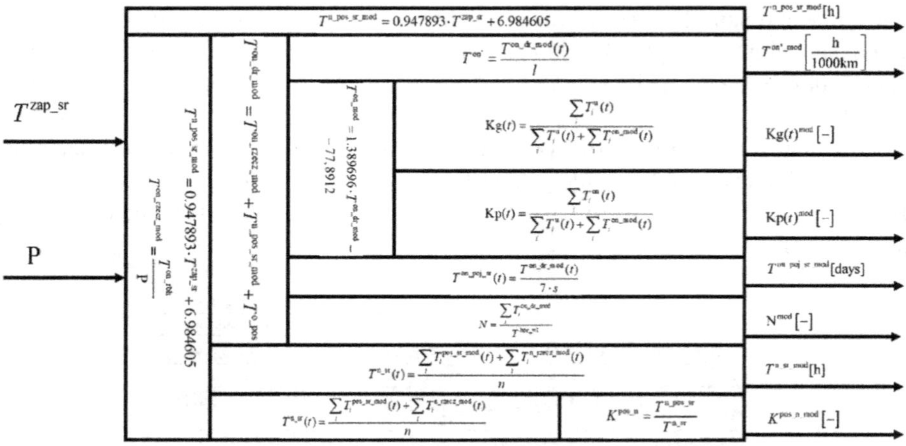

Fig. 3. The model of the TAB vehicle exploitation system

Value of the exploitation indicator obtained as part of the tests	0.97	7	14	21	28	35	42	49	Modeled values of the average waiting time for delivery of spare parts - T^{zap_sr}
Modeled values of the employee elevation gain at the service station P	1	0.971	0.967	0.963	0.958	0.954	0.950	0.945	
	1.2	0.975	0.971	0.966	0.962	0.958	0.953	0.949	
	1.4	0.977	0.973	0.969	0.964	0.960	0.956	0.951	
	1.6	0.979	0.975	0.970	0.966	0.962	0.958	0.953	
	1.8	0.981	0.976	0.972	0.968	0.963	0.959	0.955	
	2	0.982	0.977	0.973	0.969	0.964	0.960	0.956	

Fig. 4. Modeled values of the technical availability indicator Kg [–]

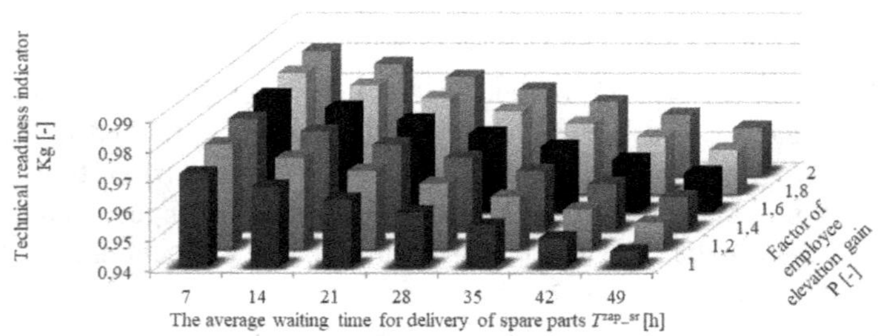

Fig. 5. Chart of modeled values of the technical availability indicator Kg [–]

4 Conclusions

The mathematical model of the Vehicle Exploitation System (VES) of the TAB and the database (TSC) analysis presented here suggest the possibilities of increasing the effectiveness of the operation of the air base (e.g. the logistics system of technical materials VES, the area of the MLS). The system of supplying technical materials (e.g. Vehicles consumables) has a significant impact on the TAB vehicle operation system, and its examination allows to determine mutual relations and formulate operational problems (e.g. The numbers of employee at the service station).

The experiment is a simulation that provides the system administrator with a depiction of the vehicle operation system resulting from changes in input parameters and allows more informed decisions. The mathematical model presented in the article, resulting from experimental research, requires thorough analysis of processes and cannot be used in the same form outside the research area. However, it is possible to create a methodology for the development of this type of models in the field of functioning operational systems (e.g. in all aviation or in the entire armed forces). The mathematical model created using the experimental method should include the largest possible research sample to capture the actual cause-and-effect relationships, not the random correlation. If it is not possible to find sufficiently high correlations between occurring phenomena (i.e. changes in one quantity do not cause changes in other quantities), then building a mathematical model without additional analysis is impossible. The developed mathematical model reflects the relationships occurring in the examined vehicle operation system and can be used as a tool supporting decision making aimed at solving operational problems.

References

1. Ben-Daya, M., Duffuaa, S., Raouf, A., Knezevic, J., Ait-Kadi, D.: Handbook of Maintenance Management and Engineering. Springer, London (2009)
2. Campbell, J.D., Reyes-Picknell, JV.: Uptime. Strategies for Excellence in Maintenance Management. Taylor & Francis Group (2016)
3. Chukowski, T.: Baza Lotnictwa Taktycznego jako element terytorialnego systemu zabezpieczenia logistycznego SZRP. WKOL AON, Warszawa (2016)
4. Chukowski, T.: Uwarunkowania funkcjonowania systemu eksploatacji pojazdów samochodowych w bazie lotnictwa taktycznego. In: Łodygowski, T., Ciałkowski, M., Pielecha, I. (eds.) Nasze stulecie. Nauka dla obronności. Wyd. Poznan, pp, 205–214 (2018)
5. Chukowski, T., Kiciński, M.: The characteristics if the vehicle maintenance system in tactical air base. J. Mech. Transp. Eng. **1**(71), 5–16 (2019)
6. Doktryna logistyczna Sił Zbrojnych Rzeczypospolitej Polskiej DD/4 (B). Sztab Generalny, 888/2014 Bydgoszcz (2014)
7. Instrukcja o zarządzaniu eksploatacją sprzętu wojskowego w Siłach Zbrojnych Rzeczypospolitej Polskiej zasady ogólne DD-4.22.13(A). IWsp SZ, Bydgoszcz (2018)
8. Instrukcji o gospodarowaniu sprzętem służby czołgowo-samochodowej DD-4.22.2. IWsp SZ, Bydgoszcz (2013)
9. Murphy, D.: Fleet Purchasing, Maintenance and Reliability. Industrial Press, Inc. Murphy, Diarmaid (2014)
10. Niewczas, A., Rymarz, J., Debicka, E.: Stages of operating vehicles with respect to operational efficiency using city buses as an example. Maint. Reliab. **1**(21), 21–27 (2019)
11. Niziński, S.: Eksploatacja obiektów technicznych. WITPiS – ITE, Sulejówek/Radom (2002).
12. Wallace, J.: Vehicle maintenance control and analysis specialist (AFSC 47254). Extension Course Institute, Air University (1986)
13. Wojciechowski, Ł., Cisowski, T.: Metoda optymalizacji procesu eksploatacji w ujęciu bezawaryjności pojazdu. Monografie. Wyd. PL, Lublin (2018)
14. Wsparcie i zabezpieczenie techniczne Sił Zbrojnych Rzeczypospolitej Polskiej. Zasady funkcjonowania DD-4.22.(A). IWsp SZ, Bydgoszcz (2017)
15. Wytyczne Komendanta 1 Regionalnej Bazy Logistycznej z dnia 28 grudnia 2016 r. w sprawie funkcjonowania podsystemu zaopatrywania technicznego w rejonie odpowiedzialności 1 Regionalnej Bazy Logistycznej (2016)

16. Wytyczne Szefa IWsp SZ z dnia 03 listopada 2017 r. w sprawie zabezpieczenia logistycznego funkcjonowania jednostek organizacyjnych Sił Zbrojnych w 2018 r. (2018)
17. Wytyczne Szefa IWsp SZ z dnia 09 grudnia 2010 r. w sprawie zasad funkcjonowania systemu zaopatrywania wojsk stacjonujących na obszarze RP (2010)
18. Wytyczne Szefa IWsp SZ z dnia 31 maja 2016 r. w sprawie funkcjonowania podsystemu technicznego Sił Zbrojnych Rzeczypospolitej Polskiej (2016)
19. Żółtowski, B., Niziński S.: Modelowanie procesów eksploatacji. ITE – PIB, Radom (2010)

Modeling the Document Control at the Airport

Natalia Czołgosz(✉)

Wroclaw University of Technology, 27 Wybrzeże Wyspiańskiego Street, 50-370 Wrocław, Poland
229466@student.pwr.edu.pl

Abstract. The increasing number of passenger flights in Europe might lead to an overflow of passengers service systems and long waiting times for the service time, and as a result of the decrease in the level of service. This paper focuses on issues related to the passenger flow in the arrival hall and in the document control areas at the destination airports, which is not widely depicted in the scientific works, but cannot be omitted in the management of airports, especially in the process of improving the airport capacity. This paper describes a method for categorizing passengers in the arrival area in terms of the Schengen Borders Code. In addition, it provides an overview of the possibilities of implementation of biometrical gates instead of traditional control desks operated by Border Guards. The article provides information about the possibility of using the simulation modeling method in that area of designing airport processes. The article is based on a case study of the document control process at the Wroclaw Airport.

Keywords: Simulation model · Level of service · Document control

1 Introduction

The increasing number of passenger flights in Europe [6, 7] has an impact on the functioning of all of the airport areas. It might lead to an overflow of passengers and long waiting times, and as a result – in a decrease in the level of service. Owing to the fact that the level of service is considered as one of the most important measures connected with general passenger satisfaction [4] that affects the income of the airport [2], there is a need of improving service processes [11]. However, the most visible effects of the lack of appropriate management with impact on the level of service might occur in the departure hall [8, 10]. Directing all of the managerial attention on that area might lead to an overload of the arrival hall. The effects of an overflow of passengers in this area might not reflect directly on the airport income, though problems with border security and passenger dissatisfaction in the last part of their travel can occur. Hence, the subprocesses in the arrival hall cannot be omitted in the management of airports, especially in the process of improving the overall airport capacity.

One of the attributes connected with passenger handling in the arrival area is categorizing passengers depending on their travel direction, not alike to the departure hall, where passengers are sorted mostly according to class. The second part of the article describes the method of categorization of passengers in the arrival area in terms of the

© The Author(s), under exclusive license to Springer Nature Switzerland AG 2021
A. Kwasiborska et al. (Eds.): ATE 2020, LNITI, pp. 11–20, 2021.
https://doi.org/10.1007/978-3-030-70924-2_2

required manner of control and identification, depending on their travel direction, to satisfy the law determined by international border agreements. It is also necessary to point out the impact of organization of space that passengers wait in and the possibilities of interruptions that might occur in the process on the total waiting time. In addition, an overview of the possibilities of implementation of biometrical gates instead of traditional control desks operated by Border Guards is given. That is the solution that might be deployed at airports, where the passenger traffic from European countries that are not under the jurisdiction of the Schengen Borders Code is relatively large.

The article also provides information about the possibility of using the simulation modeling method in that area of designing airport processes. It might be useful in the case of planning the reorganization of airport subsystems due to improving overall capacity without interference in existing infrastructure. The model described in the article was designed to examine the influence of replacing traditional control desks with biometrical gates on the total waiting time for service. The article presents the basic structure of the process, the proposal of empirical research providing data about the existing system, the algorithm of proceeding in the created model, the main input data required to describe and model the passenger flow in the check in area, and the output data provided to the user by the model. The results of the simulation can be compared with the IATA recommendations of estimated waiting time for the document control points to evaluate the proposed solutions of reorganization. The article is based on a case study of the Wroclaw Airport, a regional airport in Central Europe, which handles passengers from various groups and it is considered that the implementation of biometrical gates might shorten the total service time, increase the Level of Service and improve the capacity of the border control area.

2 The Subsystem of Check in at Arrival

All of the actions concerning passenger handling in the check in subsystem take place in the arrival hall, which is a part of a secure restricted area. The arrival hall is mostly divided into two main sections – the baggage claim area (connected with the exit hall, leading passengers to the open access area of airport) and the passport control area. The passport control area must be physically separated, as it is considered as the country border [15]. Because of that, the unwanted flow of passengers which are subject of control is indisposed. The size of the passport control area depends on the infrastructure and the presumptions being made during the airport planning stage. However, the expected passenger load might be not sufficient, if the number increases or the structure of arriving passengers changes. In that case, the overflow of passengers brings the possible dangers of extended waiting time and decreased space, which are both parts of the level of service [4].

The occurrence of that problem might be solved by expanding the arrival area through infrastructure, or minimalized with changing the organization of the processes. The improvement and the control of the processes has to be preceded by analysis of the currently working system. The most impactful aspect of check in is the structure of arriving passengers, which might be variable for a particular airport and depend on the season, current political situation and other factors. Also the methods of control being used at the airport might effect differently on the overall system.

2.1 Categorizing Passengers in the Arrival Area

As ID control must be handled in a manner that would meet the legislative requirements, the process of categorizing passengers is a necessity in the arrival area. Regulation of the European Parliament and of the Council on the Union Code on the rules governing the movement of persons across borders (the Schengen Borders Code) provides the regulations to differentiate the control process and required documents depending on the passenger citizenship and the departure airport. In European airports, there are two main categories of passengers: passengers from Schengen countries and passengers from third-countries. In the second group, in terms of the operating control subsystem, it might be useful to divide passengers into two subgroups: passengers with EU citizenship and passengers from other countries. The basic categories are presents in the Fig. 1.

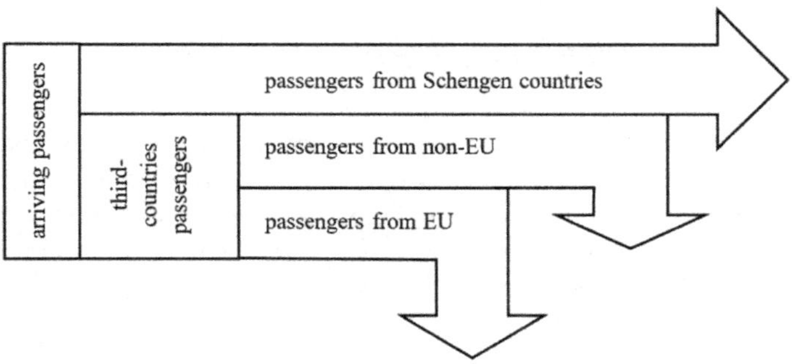

Fig. 1. Basic categorization of the arriving passengers in the system considered in the case study

The ratio of the particular group of passengers should be taken into consideration during the airport designing phase, since it determines the required space and infrastructure in this area. However, over time the ratio might change and improvement will be required.

The process of handling the passengers depends on the group they are classified to. After landing, passengers are guided to the exit of the restricted area via dedicated paths to fulfill the sequences of required steps of the arrival process. The model of the process is pictured in Fig. 2.

- Passengers from Schengen countries are guided straight to the baggage claim area. Due to Schengen regulations, there is no need to proceed to ID control in their case. Passengers receive their luggage and ID pass in the open access area.
- Passengers from third-countries are guided to the document control area. Depending on the strategy implemented in the airport, they might be split to EU and non-EU groups or be handled together. In the control area, the queue (or queues, if passengers have been divided) is formed. Passengers are consecutively served at the control points, and after ID control (ID card in case of EU passengers, passport and other required documents in case of non-EU passengers) they are allowed to get to the baggage claim area. Passengers collect baggage and pass to the open access area.

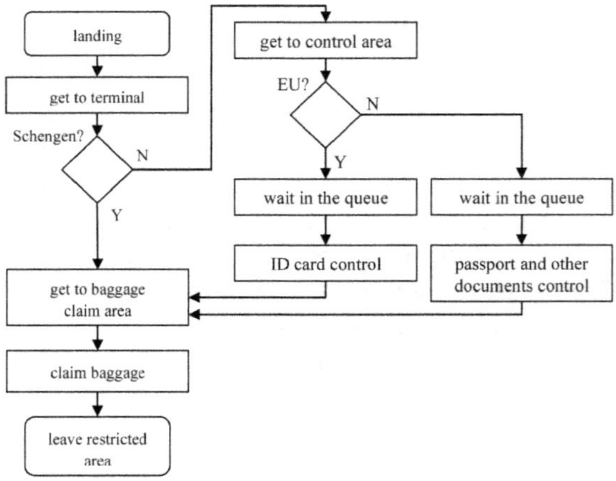

Fig. 2. Process of handling arriving passengers in the system considered in the case study

The total time passengers spend in the control area is essential for the level of service in the control process. It depends mostly on the number and the efficiency of control points, the number of passengers in each group and other factors, which might occur in real systems, e.g. disruption in queue flow, underestimated number of workers, technical problems, incorrectly projected sign and information systems etc.

2.2 Methods of Control

The essential issue of check in ID control is to assure the border security. In case of Schengen citizens, a passenger is obliged by law to submit their ID card (EU passengers), which allows to identify the person. In case of passengers from countries from beyond Europe, there is a need of identification and verification, based on passports and other required documents. When the control process is successfully completed, documents are stamped and the passenger is directed to the baggage claim area. In case of problems with identification or verification, the passenger is directed to the second line check, where the further steps of control are enforced.

Traditional border control, which is widely used in many airport systems, consists of an inspection carried out by a Border Guard. That method might be used in every case of control. Border Guard can easily proceed with the identification through visual comparison of the face of the person with the photograph in the ID documents or, if for some reason it is impossible, identification utilizing finger prints might be applied. Verification might be also proceeded by Border Guard with the use of security data base software. In case of problems, the Guard might execute further steps in second line check.

Schengen regulations permit to omit the control in case of passengers from member countries. Moreover, the recommendations for non-Schengen countries led to the unification in the documentation area and owing to that, essential biometrical data are contained in commonly used documents. With constant spreading of this kind of documents, the possibility of identification of a passenger without control proceeded by a Border Guard, when the additional verification of documents like visas etc. is not needed, is created. For automatized control, biometrical passport gates are being used. Based on face recognition systems or fingerprints recognition systems, it is possible to identify a passenger without the participation of Border Guard. The technology has been known for some time, but research in that field is in its development [13–15]. That method is permissible in accordance to Schengen regulations and might be applied to passengers from European Union. It is also a solution that decreases the number of Border Guards required to serve huge amount of passengers – it might be reduced to only one guard watching over several control points [3, 15].

The issue of using the biometrical control gates is however strictly connected with the ratio of the number of passengers in each category. They could be used as the solution for handling larger loads of people, especially in the most congested hours, with no need of increasing the number of security staff. Contrarily, in case of the need of handling large numbers of passengers demanding document verification, decreasing the number of Border Guards and traditional control desks might provide to overload and undesirable elongation of waiting time.

2.3 Factors Affecting the Control Process

The control process is defined by factors effecting on particular steps, translating into the global effectiveness of process. Main factors are connected mostly with the behavior of passengers and airport and security staff. The others relate to airport infrastructure and equipment. These include:

- the number of accessible control points,
- the daily flight schedule,
- the category of passengers ratio,
- the experience of airport and security staff,
- passengers' self-preparation to control,
- occurrences of passenger demanding second line control,
- staff's and passenger's cultural and psychophysical conditions,
- the mobility of passengers,
- the queue forming method,
- the total space of the control area,
- the technical condition of the equipment,
- the information system.

Particular airports might distinguish some other factors. Measured and analyzed, these may be useful to understand process of ID document control and point out the sectors demanding improvement or reorganization.

3 Modelling the Check in

Direct implementation of the new solutions is mostly expensive and hazardous – even advanced analysis does not provide numerical proofs of efficiency, which might be confronted with the actual performance of the system and the requirements for the new system. Therefore, in airport management, the simulation models are used for acquiring forecast data of the planned improvements, to testify and rate the efficiency of proposals. That approach is also recommended by IATA [5] and is practiced in various airport processes [6, 9, 17]. The possibility of multiplication of simulation might provide statistically significant forecasts of theoretical effects of reorganization, without interference in the working system [10].

Depending on the required type and accuracy of forecasting data, the models might be built with different approach. In the case of check in, modelling might be used e.g. to assign the appropriate number of control points, verify the efficiency of forming a queue strategy, select a proper organization of space etc. The model described in this article is focused on the issue of replacement of a number of traditional control deck with biometrical gates and its effect on the waiting time in the control area. The model was prepared for the case of airport handling of all three groups of passengers, described in Sect. 2.1, and build in the FlexSim simulation software.

3.1 Structure of the Model

The main structure of the process should be contained in the core of the model. In order to fulfill the basic functions of the document control process. The model should consider:

- the source of passengers representing the arriving passengers getting off from a landing aircraft at the time forecasted in a flight schedule,
- the queue (queues), representing the line of waiting passengers,
- the processors, representing moving from the end of the line to the control point and the proper control service,
- the sink, representing leaving the control area.

The basic idea of the model is depicted in Fig. 3. If necessary, during the preparation of the model, the introduction of more elements should be considered in order to reflect the specific conditions at a particular airport.

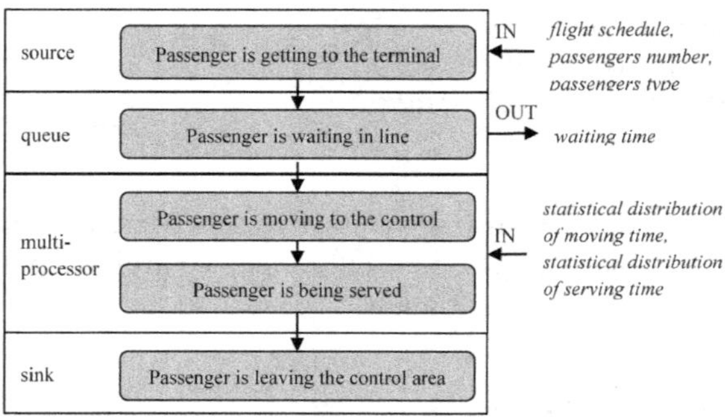

Fig. 3. FlexSim model blocks reflecting the observed subsequent basic steps of the check in process with essential input and output data marked

3.2 Input Data

In order to assure the correct forecast from the model, it is required to implement the appropriate input data, reflecting the real values describing the process. The elaborated model requires the following inputs:

- arrival schedule – number of flights and departure airport,
- number of passengers in each group in particular flight,
- number of available control points,
- estimated time of moving from the end of a queue to each control point
- estimated time of service for each group of passenger.

All the inputs are related to factors affecting the process (Sect. 2.3). Implemented values should be derived from actual conditions that might occur at a particular airport. Part of input data might be provided by the airport management system, but the empirical observations of the system prevent incorrect conclusions.

4 Check in Model Validation – Wroclaw Airport Case Study

Wroclaw Airport is one of Polish airports, handling passengers from each passenger group. Because of the increasing number of flights from Ukraine, non-Schengen countries from EU (the model was built before the UK left the EU) and charter flights, the issue of reorganization or improvement of the arrival control area was justified. The process in peak hours is inefficient, and the waiting time for many passengers is unacceptable from the perspective of the IATA Level of Service recommendation.

4.1 Local Process

After the landing, passengers are directed to the control area, where the queue (or queues in case of flights from Ukraine and Georgia) is formed. Despite eight traditional check in desks being placed in that area, in majority of cases only four of them are used because of the reduced number of Border Guards. When a flight from Ukraine or Georgia arrives, the priority for serving EU passengers for some of the desks is given. Airport staff controls the flow of passengers to particular control points.

The proposal of improvement for Wroclaw Airport was exchange of some of the check in desks with biometrical gates. Due of the relatively big number of non-EU passengers, it was decided to test the solution with three biometric gates and four check in desks.

4.2 Model Building

Observations were carried out for fifteen days in July 2019. During observations, passengers were divided into three main groups – passengers from UK (556 records of observations), passengers from charter flights (521 records of observations) and passengers from other countries (409 records of observation).

Table 1. Estimated times of service and moving to the control desk for each group (all of the distributions were described as negbinominal function)

Time of	Flights	Attributes of function	Time of	Flights	Attributes of function
Moving to control desk	UK	(2.00, 24.60)	Service	UK	(2.00, 6.88)
	Charters	(4.00, 52.90)		Charters	(7.00, 28.61)
	Other (EU)	(3.00, 42.93)		Other (EU)	(5.00, 16.21)
	Other (NEU)	(2.00, 24.97)		Other (NEU)	(2.00, 2.40)

The measurement of service time and moving to the control desk involved time that each jth passenger was being controlled $t_c(j)$, depending on the group they were representing, the time of end of service $t_e(j)$ of the jth passenger and the time of start of the service $t_s(j)$ of the following passenger. Based on that, the time of moving to each control point $t_m(j)$ was calculated as follows:

$$\mathrm{t_m}(j) = \mathrm{t_s}(j) - \mathrm{t_e}(j-1) \tag{1}$$

The input data (the estimated time of moving from the end of a queue to each control point and the estimated time of service for each group of passengers) received from statistical analysis of empirical measurements (based on obtained times $t_c(j)$ and $t_m(j)$) in FlexSim statistical module ExpertSim (Table 1), was implemented into a Multiprocessor block.

In order to simplify the model, some parts of the process were presumed:

- the arrival schedule was predefined,
- the EU/non-EU ratio was constant and defined as 2:3,
- the number of accessible control points and priorities in serving was constant during simulation,
- the time of service for biometrical gate was predefined as negbinominal (10.00, 38.46)

These issues might be measured and statistically analyzed to fill the model with their statistical distributions, but the described model was created for a general overview of performance of the system with biometrical gates.

4.3 Model Validation

Model results were verified through running a model reflecting the current system with five control desks (three with priority for EU passengers) available. The model ran a hundred times in each version and every repetition represented 24 h. The correctness of the model was justified with the test of significance for two means with the big population.

The average total waiting time for each group was tested for three variants, depicted in Table 2.

Table 2. Expected average total waiting time received from model

Number of traditional desks	Number of biometrical gates	Average time of waiting (EU) [min]	Average time of waiting (non-EU) [min]
5	0	9,2	22,7
5	2	5,4	18,9
4	3	5,5	19,8

The second variant was tested to verify the response of the model to the implemented changes. It is the most efficient alternative from those tested, but requires additional guards. The 3rd tested variant led to the conclusion that in the conditions specified in the model, it would be justified to replace four traditional desks with three biometrical gates to reach the desirable level of service.

5 Conclusions

The simulation modelling might be useful in designing many airport processes – and as it was demonstrated, also in the arrival document control area. The correctly performed research in the actual process provided specific input data that ensure the validity of the simulation. The proposed model can be used for fore-casting the effects of planned improvement.

It is important to increase the efficiency of the terminal in a balanced way, in every sector of the airport. The main efforts related to increasing the capacity of airports are focused on the departure area, although the arrival area also has to be considered as a significant part of the terminal, which demands reasonable management, especially in terms of border security. The technology of biometrical gates might be useful to increase the efficiency of this area, but their implementation should be preceded with analyses of passengers' category ratio for a particular airport.

References

1. Appelt, S., Batta, R., Lin, L., Drury, C.: Simulation of passenger check-in at a medium-sized US airport. In: Proceedings of the 2007 Winter Simulation Conference, pp. 1252–1260 (2007)
2. DKMA: Four steps to a great passenger experience (without rebuilding the terminal). https://www.dkma.com/en/images/downloads/processes/White%20paper%20-%204%20steps%20to%20a%20great%20passenger%20experience.pdf. Accessed 4 Apr 2020
3. Gunnebo: Airport security solutions. https://www.ribaproductselector.com/Docs/0/21520/external/COL690646.pdf. Accessed 4 Apr 2020
4. IATA: Airport Capacity and Level of Service. https://www.iata.org/contentassets/d1d4d535bf1c4ba695f43e9beff8294f/airport-capacity-and-level-of-service.pdf. Accessed 11 Mar 2020
5. IATA: Airport Development Reference Manual, 10th edn., pp. 183–196 (2014)
6. IATA: World Air Transport Statistics 2019. https://www.iata.org/contentassets/a686ff624550453e8bf0c9b3f7f0ab26/wats-2019-mediakit.pdf. Accessed 6 Apr 2020

7. ICAO: The World of Air Transport in 2018. https://www.icao.int/annual-report-2018/Pages/the-world-of-air-transport-in-2018.aspx. Accessed 6 Apr 2020
8. Kierzkowski, A.: Metodyka modelowania strumieni pasażerów w porcie lotniczym z uwzględnieniem wydajności, bezpieczeństwa i poziomu obsługi, 1st edn. Oficyna Wydawnicza Politechniki Wrocławskiej, Wrocław (2018)
9. Kierzkowski, A., Kisiel, T., Pawlak, M.: Model obsługi pasażerów w porcie lotniczym ze szczególnym uwzględnieniem odprawy biletowo-bagażowej i kontroli bezpieczeństwa. Prace Naukowe Politechniki Warszawskiej. Transport **122**, 39–47 (2018)
10. Kierzkowski, A., Kisiel, T., Pawlak, M.: Passenger level of service estimation model for queuing systems at the airport. Arch. Transp. **47**(3), 29–38 (2018)
11. Manataki, I.E., Zografos, K.G.: Assessing airport terminal performance using a system dynamics model. J. Air Transp. Manage. **16**, 86–93 (2010)
12. Neufville, R.: Designing airport passenger buildings for the 21st century. Transport **111**(2), 97–104 (1995)
13. Patel V.: Airport Passenger Processing Technology: A Biometric Airport Journey. Dissertations and Theses. 385 (2018)
14. Rio, J.S., Moctezuma, D., Conde, C., Diego, I.M., Cabello, E.: Automated border control e-gates and facial recognition systems. Comput. Secur. **62**, 49–72 (2016)
15. Regulation (EU) 2016/399 of the European Parliament and of the Council on a Union Code on the rules governing the movement of persons across borders (2016)
16. Transportation Security Administration. Biometric roadmap for aviation security & the passenger experience (2018). https://www.tsa.gov/sites/default/files/tsa_biometrics_roadmap.pdf. Accessed 8 Apr 2020
17. Van Dijk, N., van der Sluis, E.: Check-in computation and optimization by simulation and IP in combination. Eur. J. Oper. Res. **171**, 1152–1168 (2006)

Competitiveness Assessment of Polish Regional Airports Based on Location Planning Models

Jakub Dyrcz[1(✉)], Allan Nõmmik[2,3], Wioletta Binkowska[4], and Dago Antov[2]

[1] Politechnika Krakowska, ul. Warszawska 24, 31-155 Kraków, Poland
jakub.dyrcz.pk@gmail.com
[2] Tallinn University of Technology, Ehitajate tee 5, 19086 Tallinn, Estonia
allan.nommik@eava.ee, dago.antov@taltech.ee
[3] Estonian Aviation Academy, Lennu 40, Reola, 61707 Tartu County, Estonia
[4] Jagiellonian University, ul. Gołębia 24, 31-007 Kraków, Poland
wioletta.binkowska@doctoral.uj.edu.pl

Abstract. Airports play an important role in the whole transport system of each country. After the liberalization of the air transport market, the importance of airports, particularly in the regions, started to grow because, besides their main functions of increasing mobility, they were also considered as elements increasing economic competitiveness of these regions. With the aim to support the development of regional airports in particular, a number of investment projects was realized. Unfortunately, not all of them brought the expected success. The main goal of this article is to assess regional airports as competitors, struggling to gain and retain more passengers and, as a result, in most cases also to get additional financial support. To realize this aim, an approach to estimate the potentials of regional airports and their usage was created and, furthermore, tested based on the Polish case study.

Keywords: Regional airport · Location planning · Competitiveness

1 Introduction

Air transportation in Europe started to develop significantly after the liberalization of the air transport market. Great attention was paid to deregulation of the market. Before the liberalization of the market, domestic airports were under a number of restrictions involving regular routes abroad. That is the reason why such regional airports concentrated on feeder flights to the main hub, internal flights as a point-to-point transit and/or charter flights.

The state-owned airlines with the status of "flag carriers" obtained subsidies (Button, 2001) and had an exclusive right to serve international regular routes. On the other hand, such carriers had an obligation to organize and to operate flights to domestic airports within their own countries.

The important outcome of the market liberalization is the growing networks of state-owned airports in different countries. Airports start to be considered not only as an

© The Author(s), under exclusive license to Springer Nature Switzerland AG 2021
A. Kwasiborska et al. (Eds.): ATE 2020, LNITI, pp. 21–33, 2021.
https://doi.org/10.1007/978-3-030-70924-2_3

element of infrastructure, which guarantees mobility of the population, but also as a way to improve the economic competitiveness of regions, where they are located. But the liberalization process also has another side: some of the regional airports start to lose their traffic. This fact has already interfered with the realization of some projects in full force. The Ciudad Real Central Airport and the Gdynia-Kosakowo Airport are good examples: the first one was fully closed a few years after opening [1], the second one will not be opened for passenger transportation while the plan was to operate flights on regular routes [22]. Another case is the Castellón-Costa Azahar Airport opened in 2011, which did not realize its potential [10].

5 years ago, the European Commission published Guidelines on State aid to airports and airlines [7]. The main aim of this document is to regulate public investments in European airports, which must give much more opportunities for private capital to invest in the renovation and development of airports. Taking into account the competitive market environment, it implies that some of the regional airports could lose support from state or local authorities, which can lead to the full breakdown of their operation. However, an appropriate method to support such decision-making process is required. This research paper has two aims. The first one is to look at regional airports as competitors trying to win more passengers and, in many cases, also to get funding. And, taking into account the first aim, the second aim is to create an approach for assessment of competitors' potential as well as its usage. This article could be considered as further development of the methodology described by one of the authors in 2017 [12].

2 Definition and Classification of Regional Airports

Most of the research papers concerning regional airports focus on their concept in itself. This could be related to the fact that such airports are connected with the productivity of a certain geographic area while some of the studies accept that such a topic includes a number of problems [3].

The definition of a regional airport could be presented from different points of view. The first part of the research studies associates such airports with regional aircrafts as well as with aircrafts operating regional feeder flights. The second part of works considers the regional airport as an element of a "multiple airport region" (also called MAR). Such type of studies concentrates on regions with a number of airports. But it must be mentioned that all these approaches do not exclude the necessity to study airports individually [11].

In general, the regional airport is defined as a non-hub airport, which has no transfer traffic. Theoretically, when there are two different routes at an airport, a possibility to connect them exists. From the practical point of view, the policy of airline companies must support creating an appropriate timetable and, therefore, improvement of transfer traffic. A good example of such a situation could be the Gothenburg Landvetter airport: it served 6807631 passengers in 2018 [16], but there is no airline that would consider it as a hub for its transfer traffic. To summarize, the following definition of regional airports will be taken into account in this research paper: it is an airport that is not a hub for any airline company.

The methodology of US Federal Aviation Administration (US FAA) to classify commercial service airports is based on the amount of passengers which board a plane at a particular airport – also called enplaned passengers [21]. Such information are extracted from statistic data. There are two interesting aspects concerning the qualification of an airport for funding: (1) an airport must have regular airline traffic and (2) the number of enplaned passengers in the most cases is calculated as a percentage of actual annual passenger boardings in the whole country. Detailed information together with the application of this approach for the European environment is presented in Table 1. This calculation was done for the year 2018: it was assumed that the amount of enplaned passengers corresponds to the amount of served passengers [5].

Table 1. US FAA approach for categorization of airports in the European environment.

Annual number of enplaned passengers	Common name	European airports[(a)(b)]
Large: 1% or more	Large Hub Primary	More than 11.06 mln
Medium: at least 0.25%, but less than 1%	Medium Hub Primary	At least 2.77 mln, but less than 11.06 mln
Small: at least 0.05%, but less than 0.25%	Small Hub Primary	At least 0.55 mln, but less than 2.77 mln
Non-hub: more than 10000, but less than 0.05%	Non-hub Primary	More than 16000, but less than 0.55 mln
Non-hub: more than 2500, but less than 10000	Non-primary	More than 4000, but less than 16000

Source: own work based on US FAA, 2018. [(a)]number of served passengers in 2018; [(b)]number was adapted taking into account the population rate of the European Union

Table 2. Investment rates according to airport size.

Size of airport[(a)]	Maximum investment aid intensity
>3–5 million	Up to 25%
1–3 million	Up to 50%
<1 million	Up to 75%

Source: own work based on Guidelines, 2014. [(a)]Size is based on average passenger traffic (passengers per year)

The European Union methodology defines the limits which allow to classify all airports depending on their passenger turnover. These limits have a direct influence on decisions concerning the coverage of operating costs and the investment in airports from public funds. This approach determines investment rates and relative financial viabilities for each airport according to their size. The first parameter is destined to determine the amount of funding to support the investments in airport infrastructure, the second one - to cover operating costs. Both these parameters are elaborately described in Tables 2 and 3, respectively. Thereby, the European Union made a great step to set up and to develop the network of airports in Europe. EU countries obtained an opportunity to organize/re-organize regional airports and, therefore, to support the development of their regions. On the other hand, the main disadvantage of such EU approach is so that regional airports count on public support instead of searching for opportunities to attain self-sufficiency.

Table 3. Financial viability of airports depending on passenger traffic.

Annual passenger traffic	Description
Up to 200000	May not cover their operating costs to a large extent
Between 200000 and 700000	May not be able to cover their operating costs to a substantial extent
Between 700000 to 1 million	Should in general be able to cover their operating costs to a greater extent
Between 1 to 3 million	Should, on average, be able to cover the majority of their operating costs
Above 3 million	Usually profitable at operating level and should be able to cover their operating costs

Source: own work based on Guidelines, 2014.

Comparing US and EU approaches, it is easy to notice how they differ. The US legislation establishes very clear the bottom limit of 2500 enplaned passengers for an airport to be included in the federal funding program. In addition to that fact, there is no upper limit, crossing which will exclude a commercial service airport from support because of its profitability. The situation is quite different in EU regulations. First of all, there are no bottom limits: even if an airport serves a very low number of passengers during the calendar year, it could ask for financial support with the aim to develop its infrastructure or to cover operating costs. Secondly, there are top restrictions. It means that according to the law, if passenger traffic in the airport reaches 3 million persons, that airport cannot apply for funding for operating costs. Furthermore, if annual passenger turnover goes beyond 5 million, an airport also cannot apply for support for infrastructure investments.

3 Regional Airports in Poland

The network of regional airports in Poland is the research object of this article. The Civil Aviation Authority (Polish: Urząd Lotnictwa Cywilnego) of the Republic of Poland is responsible for maintaining the register of all airports. By the end of July 2019, the network includes 64 civil airports [17]:

- 14 certificated public-use airports,
- 9 public-use airports with limited certification,
- 21 public-use airports which cannot be subject to certification and
- 20 airports for exclusive use.

For the purposes of this research paper, only airports that are able to handle domestic and international flights were chosen. Thus, 13 certificated public-use airports and 1 public-use airport with limited certification were included in the analysis. Among airports of the first category, only the Warsaw Chopin Airport was not taken into consideration because of its status as the main airport in the country. Also it is the main interchange hub for the national "flag carrier" - Polskie Linie Lotnicze LOT S.A. From airports of the second category, the Warsaw Radom Airport was taken into account for the analysis. The reason was that in the years 2016 and 2017 it functioned normally, handling domestic and international flights. After filing for bankruptcy in 2018 this airport was sold to the State Enterprise "Porty Lotnicze" - a company that, until that moment, had already owned and managed other two airports: the Warsaw Chopin Airport and the Zielona Góra Airport. As a consequence of this transaction, there is a plan to follow the development of the Warsaw Radom Airport in the future: the first flights after the break must start in 2020 [15]. Additionally, this airport has sufficiently developed infrastructures. All this means that it will likely return to its functions as a regional airport and will be a potential competitor to other regional airports.

Fig. 1. Location of Polish voivodeships, their capitals and regional airports.

Figure 1 presents the location of all the selected regional airports in Poland (marked with letters) together with the capitals of all voivodeships (high-level administrative unit in Poland). It can be observed that the structure of regional airports' network in Poland is similar to Finland's and can be more logically compared, for example, to those of Norway or Sweden: airports do not concentrate densely in one area leaving the rest of the country "empty"; furthermore, they are spaced regularly covering the entire country.

Official statistical data concerning the number of passengers served in Polish regional airports during the period 2016–2018 is presented in Table 4. Information about passenger traffic provided by the Civil Aviation Authority of the Republic of Poland [20] shows a very clear slow trend: passenger volumes growing in all presented airports including the Warsaw Radom Airport until its closure in 2018. The biggest part of passengers in regular traffic use air transport for international trips. In 2018, the total amount of passengers traveling to and from Poland was 36 465 687 [18], which made up more than 90% of all air transport clients in the country [20]. 6 regional airports served more than 1 million passengers per year. The highest volume - 17 737 231 served passengers in 2018, which is only around 39% of total traffic – was at the Warsaw Chopin Airport, which was excluded from the analysis because its leading role in Poland. It is a relatively small number for the main country airport in comparison to other countries: for example, in Finland, the Helsinki-Vantaa International Airport handles more than 80% of total passenger traffic per year. This fact additionally confirms the existence of good opportunities for Polish regional airports to develop themselves.

Table 4. Passenger traffic in Polish regional airports, 2016–2018.

Name of airports	IATA codes	Amount of passengers			Average amount of passengers
		2016	2017	2018	
Gdańsk Lech Wałęsa Airport	GDN	3 986 410	4 601 982	4 966 949	4 518 447
Solidarity Szczecin–Goleniów Airport	SZZ	467 437	578 520	598 663	548 207
Bydgoszcz Ignacy Jan Paderewski Airport	BZG	322 135	318 400	398 066	346 200
Poznań–Ławica Henryk Wieniawski Airport	POZ	1 689 200	1 842 660	2 465 418	1 999 093
Warsaw Modlin Airport	WMI	2 859 191	2 931 503	3 080 699	2 957 131
Zielona Góra Airport	IEG	8 745	17 128	21 269	15 714
Łódź Władysław Reymont Airport	LCJ	241 256	207 377	217 426	222 020
Lublin Airport	LUZ	376 755	429 164	454 103	420 007
Copernicus Airport Wrocław	WRO	2 371 621	2 805 888	3 293 948	2 823 819
Katowice Airport	KTW	3 201 654	3 877 235	4 825 845	3 968 245
Rzeszów–Jasionka Airport	RZE	662 024	691 708	769 475	707 736
Kraków John Paul II International Airport	KRK	4 974 676	5 829 190	6 759 683	5 854 516

(*continued*)

Table 4. (*continued*)

Name of airports	IATA codes	Amount of passengers			Average amount of passengers
		2016	2017	2018	
Warsaw Radom Airport	RDO	8 965	9 903	0	6 289
Olsztyn-Mazury Airport	SZY	41 290	101 306	117 102	86 566

Source: own work based on ULC, 2019.

Some facts must be mentioned taking into account US and EU regulations. On the one hand, all selected airports meet the condition of the bottom limit described in the US law. Furthermore, 13 of them belong to the "Primary" subcategory. On the other hand, looking at the average amount of passengers from the years 2016–2018 in Table 4, there is one airport at which passenger traffic was above 5 million, and also two airports with a passenger volume of more than 3 million. According the EU rules, it must restrict their state financial support. Nonetheless, these 3 airports were left for the analysis, because they meet the declared definition of regional airports and they still compete for passengers with the other airports considered in the analysis.

4 Location Planning Models as a Tool to Assess the Competitiveness of Regional Airports

The location of an airport is an important factor influencing its competitiveness: a good position on the map near densely populated areas means proximity to potential clients, which creates a chance to obtain a significant passenger turnover. Therefore, the choice of an appropriate location to install air transport infrastructure became the research question of numerous studies. It must be mentioned here that the majority of such works was focused on finding the most suitable place, where a new airport could be built. The first research on this topic date back to 1982 [14]. During the last decade, the issue of optimization of existing networks of airports became a more and more important research direction for this field. In such a case, the facility location models could be a good tool to assess competing airports based on their map positions. Unfortunately, there is not a lot of research works which developed this idea. Within this context, the article concerning the assessment of regional airports' competitiveness in Finland using location planning models can be named as an example [12]. As it was mentioned above, the current research work presented here could be considered as further development of the methodology which was presented there.

There are two basic types of facility location models depending on the optimization aim: minimization of maximum distance and minimization of total or average distance [4]. This calculation is based on the distance matrix connecting potential locations of facilities and clients and is measured as length or travel time. The P-Center model represents the first type. Its goal is to minimize the maximum distance from each client to the closest open facility. The approach of the second type is the P-Median model. This model minimizes the sum of all distances form all clients to their nearest open facilities.

Letter P in both methods means the number of facilities, which must be included among potential ones. Also it must be mentioned that both these approaches can be calculated either in weighted or unweighted versions. The weighted models take into account the demand of clients [8].

The adaptation of both models for the research purposes of this work will be done in the same way as it was realized in the previous study of the Finnish case [12]. The chosen regional airports will be regarded as facilities which could be opened. Capitals of Polish voivodeships will be considered as clients. Populations of voivodeships will be regarded as weight parameters of clients. The important assumption is so that capitals are considered as gravitation centers of their administrative units. Taking into account previous experience, only models weighted by population were considered and calculated. This choice is based on their appropriate correlation coefficients.

Table 5. Capitals and population of voivodeships.

Name of voivodeships	Capital(s) of voivodeships	Population [persons]
Lower Silesian Voivodeship	Wroclaw	2901225
Kuyavian-Pomeranian Voivodeship	Bydgoszcz and Torun	2077775
Lublin Voivodeship	Lublin	2117619
Lubusz Voivodeship	Gorzow Wielkopolski and Zielona Gora	1014548
Lodz Voivodeship	Lodz	2466322
Lesser Poland Voivodeship	Krakow	3400577
Masovian Voivodeship	Warszawa	5403412
Opole Voivodeship	Opole	986506
Subcarpathian Voivodeship	Rzeszow	2129015
Podlaskie Voivodeship	Bialystok	1181533
Pomeranian Voivodeship	Gdansk	2333523
Silesian Voivodeship	Katowice	4533565
Swiętokrzyskie Voivodeship	Kielce	1241546
Warmian-Masurian Voivodeship	Olsztyn	1428983
Greater Poland Voivodeship	Poznan	3493969
West Pomeranian Voivodeship	Szczecin	1701030

Source: own work based on GUS, 2019.

Nowadays, Poland is divided into 16 administrative voivodeships, which are presented in Table 5 together with their capitals and populations. Additionally, two important assumptions will be taken into consideration:

- Warszawa will be considered regardless of the fact that the Warsaw Chopin Airport is situated inside the city, because there are 2 other airports (the Warsaw Modlin Airport and the Warsaw Radom Airport), which are partly or fully orientated towards the Masovian Voivodeship.
- The Kuyavian-Pomeranian Voivodeship as well as the Lubusz Voivodeship have two capitals, which means two gravitation points per administrative unit. That is why their populations were divided proportionately with the inhabitants of their capitals: Torun is a gravitation center for 760277 persons, Bydgoszcz - for 1317498 persons, Gorzow Wielkopolski - for 475834 persons, Zielona Gora - for 538714 persons.

5 Results of the Polish Case Study

The P-Center model as well as the P-median model were calculated for P equal from 1 to 14 taking into account the populations of regions. By each iteration, it was recorded how many regions were assigned to selected regional airports. The numbers of assigned voivodeships calculated at all steps were summarized in case of each airport, which finally corresponded to the number of points. The ranking of regional airports for each model was generated in compliance with these values. Basic ranking of airports was done based on their passenger traffic volumes. The results of calculations are shown in Table 6. The objective here is to assess how airports transferred their competitive potential provided by their location in winning the passengers by comparing rankings based on the calculation of models with rankings based on passenger traffic.

The Spearman's rank order correlation coefficient was used to assess and to confirm the relationship between the two variables [23]: passenger traffic volume and points obtained by calculation of both models. The choice of such approach was dictated by the research aim: both variables are a basis to create rankings of regional airports, which should help to assess their competitive potentials relative to each other as well as the usage degree of theses potentials. Calculated coefficient values are 0.398 and 0.569 for P-Center- and P-median-models, respectively. The first value shows a weak correlation, the value for the second model points out a moderate correlation. Also the null hypothesis that there is no correlation between the two variables was tested for both models separately. The significance level of the two-tailed test was set to 5% with the critical value for correlation coefficient equal to 0.538. In such a case, the null hypothesis for the P-Center-model must be accepted meaning a lack of correlation, and for the P-Median-model it must be rejected and an alternative hypothesis should be assumed, indicating the statistical significance of the relationship between the variables (the correlation coefficient differs significantly from zero).

Taking into account that the statistical significance of rank correlation for the P-Center-model does not allow to work further with this model, it was decided to calculate an aggregate ranking based on both models (the last two columns in Table 6). The correlation coefficient for such a case obtained the value of 0.609, which means higher correlation in comparison with the P-median-model alone. The significance level wasthe same. Thus, the null hypothesis was rejected, and the alternative one was accepted, which confirmed the statistical significance of the correlation between chosen variables. Further analysis was carried out based on the aggregate calculation.

The airports, the rankings of which differ the most, are the Katowice Airport, the Rzeszów–Jasionka Airport, the Łódź Władysław Reymont Airport and the Warsaw Radom Airport. According to the aggregate calculation, airports near Katowice and Rzeszow appeared in the second half of the list and took surprisingly low positions - 8 and 12, respectively - in comparison to the amount of passenger traffic served. In recent years, the Katowice Airport has maintained the third place in terms of served passengers among all regional airports in Poland. This success is associated with well-timed and good cooperation with low-cost and charter airlines, which was started in the beginning of the 2000s, when the process of Poland's acceptance into the European Union was being finished. At the moment, this regional airport has the richest offers of holiday charter flights among all competitors, which broadens its catchment area a lot. The Rzeszów–Jasionka Airport is a dynamically developing regional airport. The key to success could be the fruitful collaboration with the national "flag carrier". Besides typical feeder flights to the hub in Warsaw, Polskie Linie Lotnicze LOT S.A. operates direct connections to the USA and Israel. As an additional factor, regular flights to the Frankfurt am Main Airport organized by Lufthansa within its spoke-hub distribution system can be considered. Thus, in spite of their "worse" location according to the calculation, but thanks to their management system, airports in Katowice and in Rzeszow could find the way to improve their performance.

Table 6. Results of model calculations.

Airports	Basic ranking	Passenger volume	p-center		p-median		Aggregate calculation	
			Points	Ranking	Points	Ranking	Points	Ranking
KRK	1	5 854 516	66	1	45	1	111	1
GDN	2	4 518 447	11	9	28	5	39	5
KTW	3	3 968 245	16	6	3	12	19	8
WMI	4	2 957 131	44	2	33	4	77	2
WRO	5	2 823 819	12	8	20	6	32	6
POZ	6	1 999 093	18	4	37	3	55	4
RZE	7	707 736	5	11	5	10	10	12
SZZ	8	548 207	2	12	13	7	15	10
LUZ	9	420 007	3	13	9	9	12	11
BZG	10	346 200	17	5	12	8	29	7
LCJ	11	222 020	35	3	39	2	74	3
SZY	12	86 566	1	14	4	11	5	14
IEG	13	15 714	8	10	2	13/14	10	13
RDO	14	6 289	14	7	2	13/14	16	9

Source: own work

The situation of the Łódź Władysław Reymont Airport and Warsaw Radom Airport looks quite different. Regardless of their good locations (3rd and 9th place on the ranking in terms of aggregate calculation, respectively), they have not made a great progress to increase their passenger turnover in recent years. To the above-described problems of the Warsaw Radom Airport, it must be added that it is very young: only in 2014 it got the first certificate to become a public-use airport. Much more interesting is the situation of the airport in Lodz. It has an ideal location, but its main problem seems to be poor access, which is generally caused by the lack of necessary road infrastructure. This fact repels airlines from opening new links.

6 Discussion

Modelling showed quite consistent results that the majority of regional airports use the potential of their locations. An attempt was also made to explain most of the unclear situations, where inconsistencies between the rankings were the biggest.

Taking into account that the analyzed model based on the aggregate calculation has only a moderate correlation, it seems to be necessary to improve it, which should allow to spread further analysis. Looking at the evaluation of results, the number of available links in a regional airport could be the next key factor determining the volume of passengers served. The reason is that this factor was the main explanation of the discrepancy between the basic and the model rankings. In general, including the amount of links must be helpful in obtaining a better model. Also further research requires to understand if other factors, such as for example, famous tourist attractions in the catchment area of the airport, airport infrastructure characteristics, quality of land transport links, could play any role in shaping the better model. All these factors could be incorporated into the model as weight parameters.

There are also two other impact aspects, which could be considered to improve the understanding of the competitiveness issue. The first one is economic conditions in the country that could influence the success of regional airports. It could be, for example, GDP per capita or local economy structure. The second aspect to be taken into account is the impact of airports from neighbourhood countries. In the presented case study, the system of all regional airports was considered as a closed one. Summarizing, both of these aspects must be researched separately.

7 Conclusions

This paper presents an approach to assess the competitiveness of regional airports which is based on location planning models. Data about regional airports in Poland have been used as a case study for the analysis. The goal to assess the potential of airports to attract clients and usage of this potential was realized with the help of aggregate calculation based on P-Center and P-median models. The evaluation of the results was done by means of an approach showing how selected airports used the competitive potential of their location to obtain passengers that would count towards the real amounts of their passenger traffic.

Commenting the outcome of the Polish case study, it must be mentioned that the results of application of the proposed model were not so promising compared to the Finnish case study [12]. As written above, this model requires improvement. On the other hand, the importance of this research must be assessed through the lens of the future plans of the Polish government to build the Solidarity Transport Hub (Polish: Centralny Port Komunikacyjny), which should take the most load off of the air transport system in the country. This article could be perceived as a preliminary study, where the next steps are to include the Warsaw Chopin Airport in the calculation and then to make a model where this main airport is replaced with the new planned hub. Also other assessment approaches, such as for example, partial factor productivity indexes [13] could be taken into consideration as alternative tools.

References

1. Burgen, S.: Welcome to Don Quixote airport: cost €1bn - now it could sell to China for €10,000. The Guardian. Published online: 17.07.2015 (2015). https://www.theguardian.com/world/2015/jul/17/don-quixote-airport-building-cost-1bn-sold-china-10000/. Accessed 20 July 2019
2. Button, K.: Deregulation and liberalization of European air transport markets. Innov. Eur. J. Soc. Sci. Res. **14**(3), 255–275 (2001)
3. Chiambaretto, P.: Strategic reactions of regional airports facing competition from the high-speed train: Lessons from France. Airport Manage. **7**(1), 62–70 (2013)
4. Current, J.; Daskin, M.; Schilling, D.: Discrete Network Location Models. Facility Location: Applications and Theory. Springer, Heidelberg, p. 453 (2004)
5. Eurostat. Monthly growth and share EU-28 monthly passengers carried, 24 May 2019. https://ec.europa.eu/eurostat/statistics-explained/index.php?title=File:Monthly_growth_and_share_EU-28_monthly_passengers_carried.png/. Accessed 20 July 2019
6. Główny Urząd Statystyczny (GUS). Ludność. Powierzchnia i ludność w przekroju terytorialnym w 2019 roku, 22 July 2019. https://stat.gov.pl/obszary-tematyczne/ludnosc/ludnosc/powierzchnia-i-ludnosc-w-przekroju-terytorialnym-w-2019-roku,7,16.html. Accessed 30 July 2019
7. Guidelines on State aid to airports and airlines. Official Journal of the European Communities C 99, pp. 3–34, 04 Apr 2014
8. Hakimi, S.: Optimum location of switching centers and the absolute centers and medians of a graph. Oper. Res. **12**, 450–459 (1964)
9. Hakimi, S.: Optimum location of switching centers in a communications network and some related graph theoretic problems. Oper. Res. **13**, 462–475 (1965)
10. Liu, J.: Volotea plans Marseille – Castellon service from April 2020. Routesonline. Published online: 16.07.2019 (2019). https://www.routesonline.com/news/38/airlineroute/285407/volotea-plans-marseille-castellon-service-from-april-2020/. Accessed 20 July 2019
11. O'Connor, K., Fuellhart, K.: Cities and air services: the influence of the airline industry. J. Transp. Geogr. **22**, 46–52 (2012)
12. Pashkevich, A., Nõmmik, A., Antov, D.: Competitiveness analysis of regional airports based on location planning models: the case study of Finland. In: Proceedings of 21st International Scientific Conference. Transport Means 2017. Part III, pp. 754–761 (2017
13. Pavlyuk, D.: Spatial aspects of European airports' partial factor productivity. Transp. Telecommun. **15**, 20–26 (2014)
14. Saatcioglu, O.: Mathematical programming models for airport site selection. Transp. Res. Part B **16**(6), 435–447 (1982)

15. Suwart, K.: Lotnisko w Radomiu pilnie potrzebne. Będzie doskonale skomunikowane z Warszawą. money.pl, 19 January 2018. https://www.money.pl/sekcja/kongres-590/lotnisko-w-radomiu-pilnie-potrzebne-bedzie,80,0,2418768.html/. Accessed 20 July 2019
16. Swedavia Airports. Statistik. Utförlig trafikstatistik. Passagerarfrekvens per Swedavia flygplats, 20 July 2019. https://www.swedavia.se/om-swedavia/statistik/
17. Urząd Lotnictwa Cywilnego (ULC). Rejestr lotnisk i ewidencja lądowisk, 30 July 2019. https://www.ulc.gov.pl/pl/lotniska/rejestr-lotnisk-i-ewidencja-ladowisk/. Accessed 01 Aug 2019
18. Urząd Lotnictwa Cywilnego (ULC). Statystyki wg kierunków – państwa. Liczba pasażerów obsłużonych w polskich portach lotniczych według krajów w międzynarodowym ruchu regularnym w 2017 i 2018 roku, 25 July 2019. https://www.ulc.gov.pl/_download/regulacja_rynku/statystyki/2018/4-kw-2018/wg_krajow_regularne_kw42018.pdf/. Accessed 30 July 2019
19. Urząd Lotnictwa Cywilnego (ULC). Statystyki wg portów lotniczych. Liczba obsłużonych pasażerów oraz wykonanych operacji w ruchu krajowym i międzynarodowym - regularnym i czarterowym w latach 2016–2018, 02 July 2019. https://www.ulc.gov.pl/_download/regulacja_rynku/statystyki/2018/4-kw-2018/wg_portow_lotniczych_kw42018.pdf/. Accessed 20 July 2019
20. Urząd Lotnictwa Cywilnego (ULC). Statystyki wg przewoźników. Liczba pasażerów według przewoźników obsłużonych w polskich portach lotniczych w krajowym i międzynarodowym ruchu regularnym w 2017 i 2018 roku, 25 July 2019. https://www.ulc.gov.pl/_download/regulacja_rynku/statystyki/2018/4-kw-2018/wg_przew_regularne_kw42018.pdf/. Accessed 30 July 2019
21. US Federal Aviation Administration. Airport Categories, 23 November 2018. https://www.faa.gov/airports/planning_capacity/passenger_allcargo_stats/categories/. Accessed 20 July 2019
22. Waśniewski–Ciechorski, B.: Co dalej z lotniskiem Gdynia-Kosakowo? Gazeta Bałtycka, 04 January 2018. https://gazetabaltycka.pl/promowane/co-dalej-z-lotniskiem-gdynia-kosakowo/. Accessed 20 July 2019
23. Weathington, B.L., Cunningham, C.J.L., Pittenger, D.J.: Understanding Business Research. Appendix B: Statistical Tables. Wiley, Hoboken, pp. 453–454 (2012)

Reliability of UAVs in Fire Services Operations. Tests and Measurements of Selected Parameters

Radosław Fellner[1(✉)], Maciej Zawistowski[1], and Piotr Sadowski[2]

[1] Scientific and Research Centre for Fire Protection – National Research Institute (CNBOP-PIB), Nadwiślańska 213 Street, 05-420 Józefów, Poland
{rfellner,mzawistowski}@cnbop.pl

[2] Polish Air Navigation Services Agency, Wieżowa 8 Street, 02-147 Warsaw, Poland
piotr.sadowski@pansa.pl

Abstract. Unmanned aerial vehicles (UAVs) are increasingly and boldly used by various services, including transport services. This paper presents examples of actual and potential usage of UAVs by the fire services. Key technical parameters are defined and their measurement methodology for ensuring a high level of reliability is presented. The publication compares the obtained results with the technical parameters declared by various manufacturers. The conducted analyses of scientific literature prove that there is a research gap in the area of research on the reliability of the UAVs used in rescue and fire-fighting missions. The researchers' interest mostly focuses on the practical and technological aspects of using UAVs to prevent and respond to events such as local fires, disasters or other local threats, and more specifically, to search and rescue, provide air reconnaissance or diagnose aerial situations or inspect damaged building structures, or for other engineering inspections.

Keywords: UAV · Unmanned aerial vehicles · Reliability · Risk

1 Introduction

The use of unmanned aerial vehicles by uniformed services is connected with the necessity of their usage in extreme conditions prevalent during the operations. This is particularly the case for fire brigades whose operating conditions vary widely. For a flying platform to be able to successfully fulfill the operation goals and necessary tasks of various service formations, it must be characterized by a high safety and reliability factors, and meet specific parameters necessary to perform the mission [1].

Currently, the parameters provided by manufacturers are often determined at discretion and are not verified by any institution confirming technical parameters such as hovering time or weather resistance, such as e.g. IP. The use of such devices in very difficult conditions of the rescue operation which often accompany fire brigade operation, without first checking them may lead to the loss of equipment at best and to the death of a bystander at worst.

In the following article, the authors' aim was to determine the most important parameters affecting the possibility of UAV usage in the realities of fire service operations and the proposed technical parameters measuring methods.

© The Author(s), under exclusive license to Springer Nature Switzerland AG 2021
A. Kwasiborska et al. (Eds.): ATE 2020, LNITI, pp. 34–45, 2021.
https://doi.org/10.1007/978-3-030-70924-2_4

2 Current Use of UAVs in Fire Protection

The use of UAVs during rescue operations often eliminates the risk of spreading the threat or provides reconnaissance information. Most often UAVs are used during the daily work of the State Fire Service or the Police, however, more and more often the unmanned aerial technology is used by other formations, i.e. the Voluntary Mountain Rescue Services, the Voluntary Water Rescue Services or the Voluntary Fire Brigades [2]. UAVs show their potential in the areas of flooding, exploration, technical failures and also as a climate monitoring tool.

An example of UAVs' usage by the authorities was last year's fire of the Notre Dame Cathedral (France). During the fire on April 16, 2019, Paris firefighters and police officers used DJI-type drones to track the spreading fire and accurately aim fire hoses at the source of the fire. DJI Mavic Pro and Matrice 210 drones were used, which transmitted the live aerial video/images directly to the services participating in the rescue operations. By using this technology, firefighters could obtain the precise and necessary information to extinguish flames faster.

Another example of using of a drone during a firefighting operation is the fire in a warehouse in Manasterz town in 2019 [3]. This situation is interesting and worth mentioning because a UAV with a thermal imaging camera was used for the first time in Poland. In addition, the fire took place in a storage hall of a styrofoam factory, which emitted a lot of smoke while burning, making it impossible to observe the center of the fire using standard methods. Thermal imaging camera UAV technology again allowed for precise targeting of extinguishing streams and prevented the spreading of fire to neighboring buildings.

One cannot ignore the fact that in addition to safety, reliability and durability, an unmanned aerial vehicle should support or enable the detection of dangers, and facilitate assessment of the situation. To provide that functionality, the UAV must be equipped with additional sensors, e.g. for chemicals in the air. Garbage dump fires are difficult to extinguish, because it is not always known what exactly is burning, and this prevents the accurate selection of extinguishing agents and personal protective equipment. The authors noticed the resulting problem and therefore decided to focus attention on this aspect of measuring gaseous and dust air pollutants. The issue of air quality is also important for the conditions for conducting rescue and fire-fighting activities. Measurements of the levels of harmful substances in the air are important for safety of rescue and fire-fighting operations. Thanks to the use of UAVs and dedicated equipment, they are available and useful.

The usage of UAV creates a lot of possibilities when it is not exactly identified what materials are burnt. Early detection and identification of a harmful substance helps to neutralize it faster, reduces operating costs, can reduce the duration of activities and reduces negative impact on the environment. Obtaining detailed information on air quality during local hazards that are characterized by the emission of harmful substances (dust, combustion products, gaseous pollutants, aerosols) affecting people and the environment is crucial in the context of efficient and effective tactical decision making. UAVs equipped with proper sets of sensors allow to obtain data that requires proper reading, interpretation and sometimes additional analysis and presentation of results [4].

At this point, it should be noted that the possibility of UAV usage in firefighting structures is smooth and constantly changing depending on the needs and challenges. For example, in 2020 UAVs were used to explore forest areas and search for animals [5]. During the COVID-19 pandemic, UAVs were used to broadcast messages reminding the safety principles in force [6].

3 UAV Parameters and Their Factors

The key issue for the safety of rescuers is to use equipment with increased strength (and even aspiring to achieve the level of reliability characteristic to manned aviation equipment). Manufacturers compete in declarations regarding their products. Due to the fact that manufacturers' statements are only declarations, the author's team made the attempt to create a set of UAV operational parameters tests. The literature analysis shows that there is currently no test standard specifying how to verify the technical parameters and functionality of unmanned aerial vehicles, as well as their safety factors, reliability or strength [7–10].

Emergency services, and in particular the Fire Brigades, must operate the equipment in harsh environmental conditions, significantly different from those in which drones are used in private or civil circumstances. The above services must base their work on reliable structures, and therefore those that show a high probability that the structure (system) will not fail (will perform its functions) in the assumed time of its operation and under certain conditions [11]. The literature indicates the related concept of "robustness", which can be translated as "solidity", "strength", "resistance", which means "strong and effective in all or most situations and conditions" [11]. UAVs declared by the manufacturer as designed "for the uniformed services" must have a low susceptibility to disability, i.e. a state of inability to perform the required function [12].

To achieve the level of reliability sought, product testing is often used as a tool that reduces the risk of failure or unforeseen events.

Based on the opinions of fire brigade officers and literature analysis, the following table summarizes the parameters tested using the DronesBench and AirLogger devices and the impact of these parameters on UAV performance and performance properties relevant to the UAV operator.

In order to extract parameters describing the properties of UAV measurement capabilities and indication of the performance of sub-assemblies of parts, systems, elements, UAV's architecture and functional areas have been previously analyzed [13].

A summary of the tested parameters and their impact on UAV's performance and functions is described in the Table 1.

In fact, our measurement procedures require thorough testing of the device to check whether e.g. drone weight and battery capacity comply with UAV optimal performance standards. After the parameters monitoring is finished, the device generates a protocol confirming the technical properties of the drone.

Table 1. Summary of tested parameters and their impact on UAV performance and functions.

Measurement equipment	No	Parameter	Impact on UAV performance and functions
DronesBench	1	Power consumption	Power consumption allows to determine: - the maximum current drawn from the battery and check the selection of individual components (power supplies), - flight time in given conditions (e.g. temperature), - differences in absorbed power (recorded during service inspections) may indicate a poor technical condition of the UAV
	2	DronesBench index (IDB) – FOM (figure of merit) parameter	The higher the value of the parameter, the greater the efficiency of the propulsion system, and therefore the longer the flight time (flight time is the weakness of multi-rotors - compared to other designs). Making the flight time longer, you can make UAV stay airborne longer and obtain more data with it
	3	Energy consumption per kilogram of weight	
	4	Energy used	
	5	Mass	It allows the UAV to qualify to the appropriate weight category according to aviation law (up to 5 kg, up to 25 kg, 150 kg)
	6	Thrust	It defines the maximum mass of sensors, the additional load that UAV can lift
	7	Battery capacity	It allows to determine the status of the battery and indicates when it should be disposed of, because during the usage of the battery its capacity decreases, and therefore UAV flight time decreases. Periodic testing is important in the context of conducting rescue and firefighting activities, where the time to replace the battery is limited
	8	Center of Gravity location	The center of gravity shift causes: - reduction of the maximum load weight and uneven exploitation of the propulsion system parts, which may increase the likelihood of failure, - poorer controllability (limiting movements in the opposite direction to the center of gravity shift), - un-maneuverability
AirLogger	1	Hover accuracy (GPS mode)	The higher the hovering accuracy, the easier the UAV control, no need for wind corrections, more precise indications of the location of the objects (UAV as a reference point)
	2	Flight time	The longer the flight time, the smaller the number of landings needed to replace the battery, and thus fewer interruptions, e.g. in image transmission from the place of operation
	3	Horizontal maximum velocity	Reaching the place of the incident faster
	4	Vertical rate of climb	Faster avoidance of obstacles, resistance to convection currents (e.g. during a fire), faster landing in emergency situations (the need to give way to manned aircraft and HEMS flight operations)
	5	Vertical rate of descent	
	6	Telemetry position accuracy (3-axis gyro and accelerometer parameters)	The higher the accuracy, the better the mapping of the routes traveled, which can be useful when analyzing events (logs) with the participation of UAV
	7	Meteorological conditions (air temperature, humidity, pressure)	By measuring and recording atmospheric conditions, it can be determined whether a given UAV (for which conditions of use have been specified) can be used for operations. The lower the temperature, the faster the battery will discharge, and therefore the shorter the flight time

(continued)

Table 1. (*continued*)

Measurement equipment	No	Parameter	Impact on UAV performance and functions
	8	FFT frequency	It allows detection of UAV damages and their location through frequency analysis of vibrations: - vibrations from 0 to ~30 Hz are the frame's own (and resonance) vibrations, - frequencies from 50–100 Hz show balance/imbalance of engines/propellers, - higher frequencies (300–450 Hz) allow estimating the condition of bearings in motors The appearance of UAV vibrations may cause the so-called jelly effect, which worsens the image recorded by the camera mounted onboard

Source: own study and research based on own analysis, literature review and interviews with representatives of uniformed services.

4 Research Results

4.1 Measurements Using DronesBench

As part of the statutory work, various unmanned aerial vehicles were tested using the DronesBench device, including our own design (custom3) and UAVs from professional companies such as Spartaqs, Yuneec, DJI, Parrot (Tables 2, 3 and 4).

Table 2. Comparison of parameters of selected UAV.

Parameter	Custom UAV	Parrot Thermal	Parrot Anafi	Spartaqs Mareus
Mass [g]	2714	753	349	5443
COG shift – pitch axis	2	2	N/A	−1
COG shift – roll axis	0	0	N/A	+ 1
Battery type and parameters	6S1P, 22,2 V	3S1P, 11,1 V	3S1P, 11,1 V	6S1P, 22,2 V
Propeller parameters - diameter	12	6,1	3,89	17
Propeller parameters - pitch	5,5	N/A	N/A	5,8
Maximum thrust [kg]	4,7	N/A	N/A	N/A

In the case of this test, it can be seen that the engines are more or less at a similar level of exploitation.

In this case, it can be said that the M3 engine creates the least resistance. Since the constructions supplied from one of the largest production and distribution companies, the UAVs are well adapted and balanced for private applications, the article presents the results obtained from custom UAVs created for non-specialized applications. The basic parameters of the tested UAVs are presented in Table 5.

Table 3. Comparison of engine temperatures of custom UAV before and after switching them on attempt no. 1.

Number of engine	Engine temperature before test [°C]			Mean T [°C]	Engine temperature after test [°C]			Mean T [°C]
M1	22,9	23	22,8	22,9	24,2	24,2	24,2	23,46
M2	22,8	22,8	22,8	22,8	25	25	25	23,74
M3	23	23	23	23	24,2	24,2	24,2	23,51
M4	22,8	22,8	22,8	22,8	23,4	23,4	23,4	23,06

Table 4. Comparison of engine temperatures before and after switching them on Attempt 2.

Number of engine	Engine temperature before test [°C]			Mean T [°C]	Engine temperature after test [°C]			Mean T [°C]
M1	25	25	25	22,9	27,4	27,4	27,4	25,73
M2	26,4	26,4	26,4	26,4	27,2	27,2	27,2	26,74
M3	26,7	26,7	26,7	26,7	27	27	27	26,83
M4	26,7	26,7	26,7	26,7	25,8	25,8	25,8	26,31

Table 5. Basic parameters of the tested custom UAV.

Parameter	Value	Remarks
Mass	535 g	Weight with battery
Frame dimensions	250 × 250 × 80 [mm]	Width, length, height
Battery mass	188 g	
Propeller diameter	127 mm	
Propeller pitch	3″	

The drone has been subjected to thrust tests, center of gravity location and electrical measurements, including:

- The electric power consumed by the drone,
- The maximum current consumed by the drone,
- The DBI (Figure of Merit) parameter,
- Flight time.

The measurement results obtained are presented in Table 6.

The Figs. 1, 2 and 3 show the recorded waveforms:

Table 6. List of recorded measurement results.

Parameter	Value	Unit of measure	Remarks
Attempt time (measure time)	243	[s]	
Initial battery pack voltage	12,5	[V]	
Final battery pack voltage	10,7	[V]	
Minimum current drawn from the battery pack	0,4	[A]	
Maximum current drawn from the battery pack	25,8	[A]	
Average current drawn from the battery pack	18,53	[A]	
Mean value of current drawn from the battery pack during the test	18,9	[A]	
Maximum power consumed during the test	290	[W]	
Minimum power consumed during the test	4,6	[W]	
Average power consumed during the test	205	[W]	
Mean power value consumed during the test	206	[W]	
Maximum DBI value	34,93	DBI [mN/W]	DBI - the ratio of thrust to the necessary electrical power, which represents the health indicator of the drone as a whole
Minimum DBI value	25,21	DBI [mN/W]	
Average DBI value	0	DBI [mN/W]	
Mean value of DBI	25,57	DBI [mN/W]	
Consumed energy	48077,97	[J]	
Consumed electicity	1197,032	[mAh]	

- the value of force acting in the OZ axis
- DronesBench Index (DBI) parameters
- instantaneous power consumed
- battery current and voltage

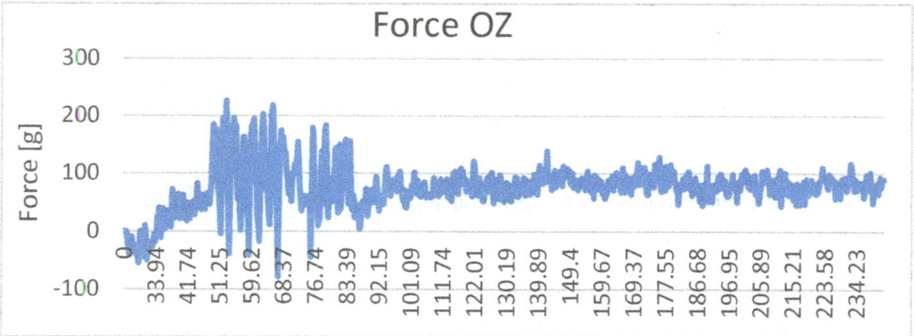

Fig. 1. Change of force in the OZ axis in time.

The maximum lift made by the drone was about 210 g.

For comparison, a longer trial was carried out, the results of which are presented below (Table 7) (Figs. 4 and 5).

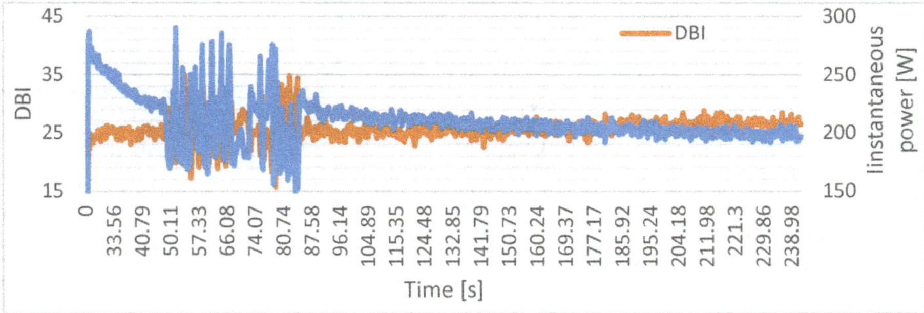

Fig. 2. Change of DBI parameter values and electric power consumed from the battery.

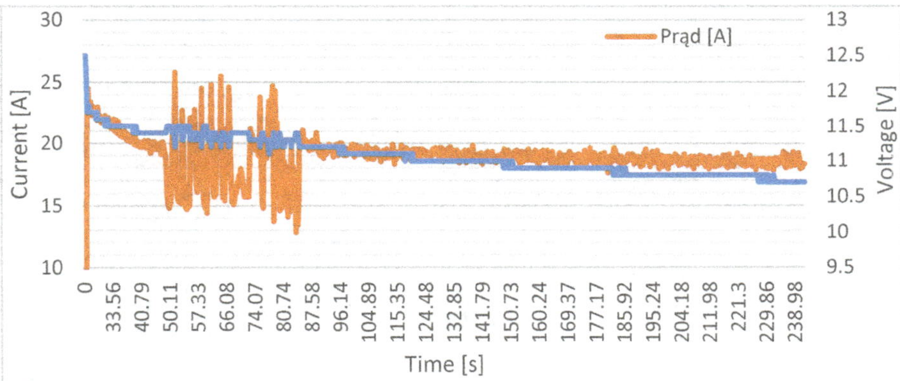

Fig. 3. Change of electrical parameters of the battery during the test.

As shown: the voltage course (blue) after the test, about 10% of its capacity was left, which in practice would result in an emergency landing. From the waveform we can observe the standard shape of the battery discharge curve, similar in shape to the theoretical waveform.

Slight deviations from the curve are caused by temporary changes in the current consumed by the UAV.

4.2 Measurements Using DronesBench

During the AirLogger measuring device UAV tests, the design team performed FFT (Fast Fourier Transform) frequency analyzes of the UAV's own vibrations. The information collected in this way can help in determining the technical condition of the machine and faster detection of the damage element. This is due to the fact that the various mechanical components that make up the UAV vibrate with different natural frequencies, e.g.:

- Vibrations from 0 to ~30 Hz are the frame's own (and resonance) vibrations,
- Frequencies from 50–100 Hz show the balance/imbalance of engines/propellers,
- Higher frequencies (300–450 Hz) allow estimating the condition of bearings in motors.

Table 7. List of recorded measurement results.

Parameter	Value	Unit of measure	Remarks
Attempt time (measure time)	378	[s]	
Initial battery pack voltage	11,4	[V]	
Final battery pack voltage	9	[V]	
Minimum current drawn from the battery pack	26,5	[A]	
Maximum current drawn from the battery pack	0,6	[A]	
Average current drawn from the battery pack	17,87	[A]	
Mean value of current drawn from the battery pack during the test	17,7	[A]	
Maximum power consumed during the test	292	[W]	
Minimum power consumed during the test	0,3	[W]	
Average power consumed during the test	190,3	[W]	
Mean power value consumed during the test	206	[W]	
Maximum DBI value	37,12	DBI [mN/W]	DBI - the ratio of thrust to the necessary electrical power, which represents the health indicator of the drone as a whole
Minimum DBI value	0	DBI [mN/W]	
Average DBI value	35,96	DBI [mN/W]	
Mean value of DBI	70871,2	DBI [mN/W]	
Consumed energy	1850,51	[J]	
Maximum DBI value	37,12	[mAh]	

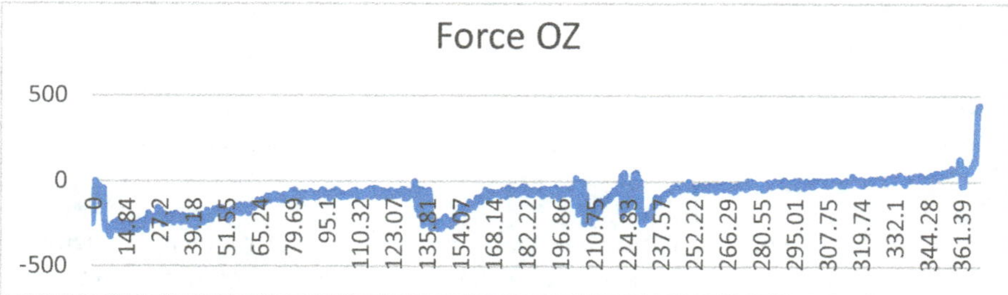

Fig. 4. Change in UAV thrust in the OZ axis over time.

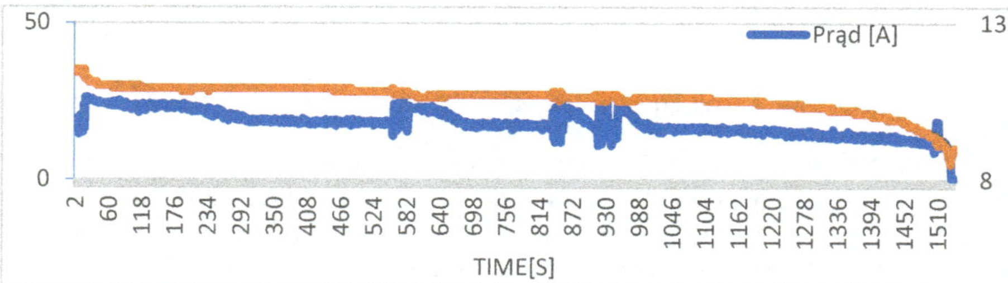

Fig. 5. Change in the electrical parameters of the battery during the test.

The tests were carried out for the UAV in hovering condition. Currently, AirLogger allows the analysis in the 0–50 Hz spectrum, and therefore only the frame's own vibration was measured. This is due to the AirLogger sampling frequency fp = 100 Hz and Nyquist frequency fNq = fp/2 = 50 Hz. Ultimately, IMU measuring systems installed in the AirLogger can collect data with the maximum frequencies:

- MPU9250, fp = 3.6 kHz;
- BNO055, fp = 1 kHz;

Interesting frequency distributions for Gx, Gy, Gz (axis) were recorded for the examined platform (Figs. 6, 7 and 8).

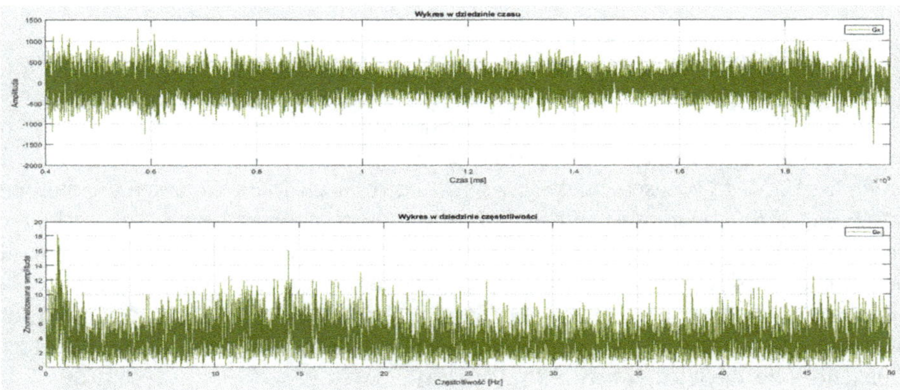

Fig. 6. Changes in angular velocity values in the OX axis in the time and frequency domain.

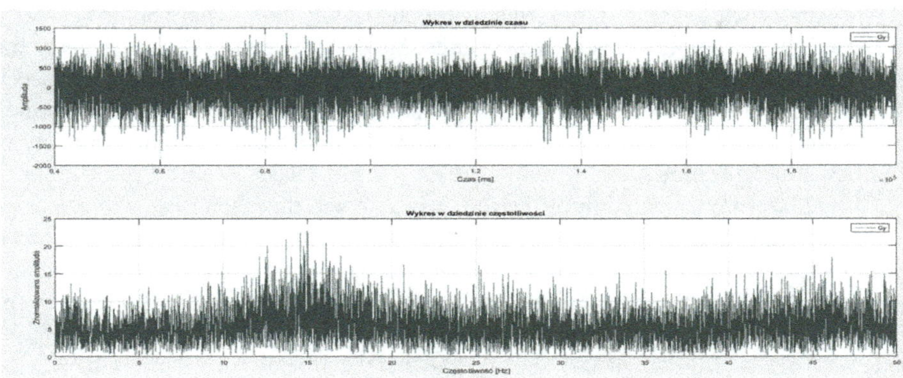

Fig. 7. Changes in angular velocity values in the OY axis in the time and frequency domain.

For the recorded distributions, a slight increase in the frequency spectrum was recorded in the OX, OY, OZ axes around 14–15 Hz. After a closer look at the structure, it was found that the frame actually has a slight looseness in the joints.

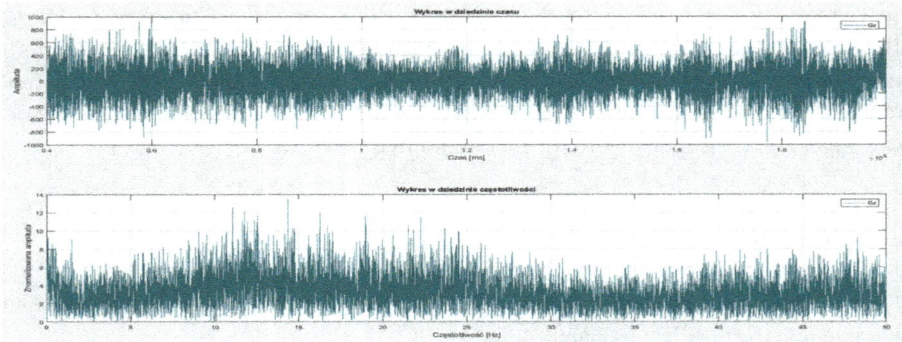

Fig. 8. Changes in angular velocity values in the OZ axis in the time and frequency domain.

5 Conclusions

UAVs and their equipment in particular are developed rapidly. This allows them to be used for various purposes - also during rescue operations as indicated in Table 1. In order for UAVs to be used safely they must be resistant to environmental conditions that occur during operations and should be tested for real parameters. The AirLogger and DronesBench devices purchased as part of the status research activity allow the determination of selected parameters, UAV operation status and suitability for specific applications. Another example of UAV application is the use of gas and ppm sensors mounted onboard to measure air pollution to recognize the place of incident, detect and then determine the level of harmful substances.

The results of this study may allow for the initial systematization of formal, legal, research and operational aspects regarding the use of UAVs during rescue and firefighting activities. Up to now, there have been no such analyzes performed. The results of this work can be developed for further research and reflection. They should be treated as part of a very complex issue, which is undoubtedly the use of UAVs in rescue or fire preventing operations. The team is aware of the contributing nature of this research paper, which in no way should be perceived as completed. The team notes that the research efforts and results obtained are still to be continued to increase reliability or safety of the UAV structures used.

During the implementation of the test, it was noted that there was still a lack of adequate methods of UAV verification used by uniformed services in suboptimal conditions (operation or mission conditions). One of the conclusions is the legitimacy of deepening the subject of research methodologies and parameters of UAVs used by the services. Quality management systems and tests determined from the perspective of the end user's needs of fire brigades is a challenge that the project team wants to face in the near future.

The UAV test results will be relevant for the services involved in the rescue and firefighting (and other) activities during the selection of UAVs they buy. The analyses, trials and tests undertaken will make it possible to find practical application in the selection of UAVs for the planned mission and activities, as well as provide the basis for further research and development works on the construction and application of UAVs in the fire service.

In the assessment of the research team, this report may be useful for, among others:

- developing a concept to reduce the risk of defects in UAVs to an acceptable level;
- developing technical recommendations and recommendations increasing the reliability and performance of UAVs;
- determining UAV parameters for the needs of future tenders for State Fire Services (and other authority services);
- conceptual work on the construction of a special UAV for fire brigades;
- developing instructions on the use of UAVs during rescue and fire-fighting operations as well as the concept of operations or rules for organizing rescue using UAVs;
- works related to UAV standardization and certification.

References

1. Feltynowski, M., Zawistowski, M.: Threats related to the use of unmanned systems in emergency services. Saf. Fire Tech. **51**(3) (2018)
2. Feltynowski, M., Zawistowski, M.: Opportunities related to the use of unmanned systems in emergency services. Saf. Fire Tech. **51**(3), 126–136 (2018)
3. A fire in a carpentry workshop in Podkarpacie. There was no way to put it out. https://www.polsatnews.pl/wiadomosc/2019-12-16/pozar-w-zakladzie-stolarskim-na-podkarpaciu-nie-bylo-szans-tego-ugasic/. Accessed 17 May 2020
4. Harmata, W., Witczak, M., Pietrzak, G.: Aerial detection of contamination with the use of unmanned vehicles–development prospects. Sci. J. Mil. Univ. Land Forces **50**, 5–24 (2018)
5. Drones above Wolsztyn's forest. https://www.springer.com/lncs. Accessed 1 May 2020
6. Głogów: Appeals to the inhabitants will flow from the drone's speakers. https://glogow.naszemiasto.pl/glogow-apele-do-mieszkancow-poplyna-z-glosnikow-drona/ar/c1-7622727. Accessed 17 May 2020
7. Fellner, R.: UAV standardization and certification in the light of ISO and EU work. In: Kochańczyk, R. (ed.) Legal and Certification Aspects of Unmanned Aerial Vehicles in the Light of Selected International Regulations, pp. 34–48. Police School in Katowice (2019)
8. Zawistowski, G., Pietrzela, D., Kaczmarzyk, P.: CNBOP-PIB research and certification potential in the field of the UAV dedicated for use by uniformed services. In: Feltynowski, M. (ed.) The Use of Unmanned Aerial Platforms in Operations for Public Safety, CNBOP-PIB (2019)
9. Tanzi, T., Apvrille, L., Dugelay, J.L., Roudier, Y.: UAVs for humanitarian missions: autonomy and reliability. In: 2014 IEEE Global Humanitarian Technology Conference, pp. 271–278. IEEE, San Jose (2014)
10. Petritoli, E., Leccese, F., Ciani, L.: Reliability and maintenance analysis of unmanned aerial vehicles. Sensors **18**(9), 3171 (2018)
11. JARUS guidelines on SORA: Annex I glossary of terms, JAR-DEL-WG6-D.04, p. 11 (2017)
12. IEC 60300-3-1 dependability management part 3-1: application guide analysis techniques for dependability guide on methodology, p. 13 (2003)
13. Valavanis, K.P., Vachtsevanos, G.J. (eds.): Handbook of Unmanned Aerial Vehicles. Springer, Dordrecht (2015)

Brexit. What Will Happen to Flights Between the United Kingdom of Great Britain and Northern Ireland and European Union Countries? Legal Aspects of Air Transport

Agnieszka Fortońska(✉)

Silesian University, Bankowa Street 12, 40-007 Katowice, Poland
afortonska@gmail.com

Abstract. On 29 March 2017 the United Kingdom of Great Britain and Northern Ireland gave notification of its intention to withdraw from the European Union in accordance with Article 50 of the Treaty on European Union. Under that Article, each State has the possibility to withdraw from the European Union. However, such a decision has a number of legal and political implications. The purpose of this paper will be to discuss issues relating to the United Kingdom's withdrawal from the European Union and the operation of flights after Brexit. The article will address the issue of the legal aspects of the operation of flights, their safety and United Kingdom's relations between aviation organisations and other countries. The main question that can be asked is how flights from the UK will be operated and whether they will still be safe. In addition, it should be considered whether, in the current legislation status, UK will be under the jurisdiction of the European Union Aviation Safety Agency or another aviation organisation.

The purpose of the presented research is:

1) To Examine the Relationship Between the European Union and to Analyse the Current Legal Situation Regarding the Operation of Flights from and to the United Kingdom;
2) To Examine the Relationship Between the European Union Aviation Safety Agency and the United Kingdom;
3) To Analyse Relationships Between the United Kingdom and Other Aviation Organisations.

The following research methods will be used in the article: dogmatic and legal, based on the examination of the content of regulations, theoretical and legal, used to examine the processes of creating regulations. It should be noted that the withdrawal of a Member State from the European Union is quite an unusual procedure, which has many legal consequences. However, to maintain the continuity of flights, the European Union has set rules that will apply to the United Kingdom after leaving the community.

(In the further part of the paper, the name United Kingdom of Great Britain and Northern Ireland will be used instead of the official name of the country. In the further part of the paper, the abbreviation EASA or the Agency will be used).

© The Author(s), under exclusive license to Springer Nature Switzerland AG 2021
A. Kwasiborska et al. (Eds.): ATE 2020, LNITI, pp. 46–56, 2021.
https://doi.org/10.1007/978-3-030-70924-2_5

Keywords: Safety · Brexit · European Union · Civil aviation

1 Introduction

The European Union is an international organisation which currently comprises 27 countries. It has its own privileges and immunities. In addition, it is entitled to make claims, but also bears responsibility under international law. It also enjoys a secondary *ius legationis.* The basis for the functioning of this community is the Treaty on European Union [1] and the Treaty on the Functioning of the European Union [2]. According to Article 3 of the Treaty on European Union, the States which belong to the Union shall be associated to promote peace, its values and the well-being of its peoples [1]. In addition, the Union shall offer its citizens an area of freedom, security and justice without internal frontiers, in which the free movement of persons is ensured in conjunction with appropriate measures with respect to external border controls, asylum, immigration and the prevention and combating of crime. In this way, citizens of the Member States can move freely between the Member States of the European Union. Anyone who holds the nationality of a Member State is also a citizen of the European Union (Article 20 TEU). This is an additional citizenship, which does not replace national citizenship, but guarantees citizens many rights, such as political ones. Moreover, such citizenship also gives rise to obligations, such as taxation. However, if a State and its citizens want to obtain many privileges from the membership in the European Union, they have to meet certain requirements. According to Article 49 TEU, any European State which respects and is committed to promoting values such as human dignity, freedom, democracy, equality and respect for human rights may apply to become a member of the Union. The European Parliament and national Parliaments shall be notified of this application. The applicant State shall address its application to the Council, which shall act unanimously after consulting the Commission and after receiving the consent of the European Parliament, which shall act by a majority of its component members. The conditions of eligibility agreed upon by the European Council shall be taken into account. It should be stressed that the conditions of admission and the ensuing adjustments to the Treaties on which the Union is founded are the subject of an agreement between the Member States and the applicant State. This agreement is subject to ratification by all Contracting States in accordance with their respective constitutional requirements [1]. Obviously, any State which is already a member of the European Union may withdraw from it under the conditions laid down in the TEU.

In accordance with Article 50, any Member State may decide to withdraw from the Union in accordance with its own constitutional requirements. A Member State which decides to withdraw shall notify the European Council of its intention. The Union shall then negotiate and conclude an agreement with that State, setting out the arrangements for its withdrawal, taking account of the framework for its future relationship with the Union. That agreement shall be concluded on behalf of the Union by the Council, acting by a qualified majority, after obtaining the consent of the European Parliament. The Treaties shall cease to apply to the State in question from the date of entry into force of the withdrawal agreement or, failing that, two years after the notification, unless the European Council, in agreement with the Member State concerned, unanimously decides to extend this period. In addition to the possibility of becoming a member of or withdrawing from the Community, there is also the possibility of re-entering the EU. This is when a country applies for re-admission and its application is subject to the same procedure as in the case of the application for membership for the first time.

2 Analysis of the United Kingdom's Relations with the European Union and the Current Legal Situation Regarding the Operation of Flights from and to the United Kingdom

For the first time in the history of the European Union, a Member State has decided to leave the Community. Since 1 February 2020, the United Kingdom has withdrawn from the European Union. This date is at the same time the date of entry into force of the Agreement on the United Kingdom's withdrawal from the Union and the first date on which the transitional period from 1 February 2020 to 31 December 2020 starts. However, before that happened, the United Kingdom had had to comply with the formal requirements under the Treaty on European Union. Therefore, on 29 March 2017, the United Kingdom gave notification of its intention to withdraw from the Union (Article 50 of the Treaty on European Union). The United Kingdom's withdrawal therefore had a number of consequences. Given the vastness of the subject, I will only concentrate on the issue of flight operations, aviation safety and the relationship between the UK and the European Union, third countries and organisations following the UK's withdrawal from the Union. It is worth noting that Regulation (EU) 2019/502 of the European Parliament and of the Council of 25 March 2019 [3] sets the conditions for the granting of the Union operating licence to air carriers and establishes the freedom to provide intra-EU air services. On the basis of this document, basic concepts such as an EU air carrier, a UK air carrier or a scheduled or non-scheduled air service were defined. Thus, already in this piece of legislation, a distinction is made between an EU and a UK air carrier as a result of the UK's withdrawal from the European Union. Under the terms of the Regulation, British air carriers will be allowed to fly across the territory of the Union without landing; make stops in the territory of the Union for non-traffic purposes and perform scheduled and non-scheduled international air transport services. In the light of the Regulation, the European Commission will monitor whether the United Kingdom discriminates against EU carriers. However, the Regulation expired on 1 February 2020, when the Agreement on the UK's withdrawal from the Union came into force [4]. It should be noted that in addition to the above mentioned Regulation, Political Declaration No. 2020/C 34/01 was concluded in 2020 between the United Kingdom and the European Union. According to this Declaration, the Parties should ensure the provision of air passenger and cargo services through a comprehensive air transport agreement (CATA). The CATA Agreement should cover market access and investment, aviation safety and security, air traffic management and provisions to ensure open and fair competition, including appropriate and relevant consumer protection requirements and social standards. Furthermore, it was emphasized that additional arrangements should be put in place by the Parties to enable them to cooperate to achieve high standards of aviation safety and security, including through close cooperation between EASA and the

Civil Aviation Authority (CAA) [5]. It is worth noting that, after the transitional period, the United Kingdom will be required to conclude: bilateral safety agreements with the European Union and other third countries in order to reach the required consensus on the mutual recognition of aviation certificates or agreements. Such an agreement will have to be concluded, e.g. with the United States, as until now the United Kingdom has been bound as a member of the European Union by an air transport agreement [6].

Another important aspect is that the withdrawal of the UK from the European Union results in the withdrawal from the European Common Aviation Area (ECAA). It should be noted that the ECAA was established in 2006 as an extension of the single aviation market and it is supervised by the European Aviation Safety Agency (EASA). Furthermore, its legislation is enforced by the European Court of Justice (ECJ). It is important that during the transitional period the United Kingdom will continue to be represented in EASA by the Civil Aviation Authority (CAA), the national aviation regulatory authority - but will no longer play a formal role in any decision-making process.

As a result, if the United Kingdom is not granted permission or new agreements are not concluded, flights from and to the United Kingdom may be suspended. As a result of the suspension of the so-called "Brexit" by the end of 2020, i.e. the end of the transitional period provided for in the agreement resulting from the withdrawal of the United Kingdom, most of the existing legislation with regard to air transport continues to apply. At the same time, the EU approved regulations 2019/494 and 2019/505 in order to secure air traffic between the UK and the EU and the European Economic Area (EEA).

However, it should be noted that EU has also negotiated horizontal agreements with 17 other countries outside the European Common Aviation Area. These agreements (including the EU-US Open Skies Agreement and the EU-Canada Air Transport Agreement) cover areas such as access rights for airlines, passenger rights and investment. As a result of the agreements and the ECAA membership, flights from the UK can operate to 44 countries. However, in order to maintain access to international flights after 31 December 2020, the UK has to take action to ensure the continuation of air travel by concluding appropriate agreements with countries. As a consequence of the UK leaving the EU, agreements with third countries will cease to apply after the transitional period. The UK will then be outside the Common European Economic Area. Also, British airlines will no longer be subject to ECJ jurisdiction or EASA supervision. In order to maintain its current relationship with countries, the UK is renegotiating its own bilateral agreement with the US, Albania, Georgia, Iceland, Israel, Kosovo, Montenegro, Morocco and Switzerland, or Canada.

Another aspect to be discussed is the UK aviation licences. It is important to underline the fact that if no agreement is reached on future relations between the UK and the EU, UK business licences will expire in the EU. Thus, UK-based airline operators will have to relocate their principal place of business to an EU Member State and hold a majority stake in the EU in order to be able to continue operations in the EU. Furthermore, all certificates, licences and registrations issued by the CAA on behalf of EASA for both pilots and aircraft and parts thereof will also cease to be valid and will have to be re-approved in an EU Member State.

In view of the above, consideration should be given to what action, in addition to the conclusion of agreements, the United Kingdom can take to maintain continuity of flights to other countries and enjoy the nine freedoms of the air. It appears that one possible solution is to obtain the status of ECAA member as Bosnia and Herzegovina has done. Then the United Kingdom, no longer being a member of the Union, would have the status of a member from outside the European Union. However, this possibility may not be accepted since, being a member of the ECAA, the United Kingdom would again be under the jurisdiction of the European Court of Justice and EASA, the institutions and agencies of the European Union. Consequently, the United Kingdom would have to submit to the European Union again, which would be contrary to the Brexit concept.

In addition, the text of the ECAA agreement itself (Article 33) requires a candidate wishing to join the ECAA: to comply with EU aviation law and to adapt legislation in the framework of close economic cooperation or "association agreement" with the EU [7]. It is important to underline that, in addition to these considerations, the consent of the current signatories to the ECAA Agreement is required. At this stage, the question arises whether all countries would agree to the UK joining the ECAA. Unfortunately, this is a debatable issue. It should be noted that the lack of acceptance of UK membership by one of the countries may be aimed at eliminating low-cost airlines from the European market. On the other hand, the consent may be dictated by the self-interest of Member States that want to have access to one of the larger aviation markets, which is the British market.

Another solution could be the conclusion of a bilateral agreement, such as the one that Switzerland has with the ECAA. However, even in this case, the supervision of the ECJ and the dispute over Gibraltar will be a matter for debate. Another viable scenario, after the end of the transitional period, could be a new agreement, between the UK and the EU, which would be similar in nature to the existing open skies agreements. It should be mentioned that another possibility is the conclusion of individual bilateral agreements with each country belonging to the European Union.

In case of a failure to reach an agreement with the European Union, it is feasible to return to the old provisions, which were laid down in the 1944 Chicago Convention. As a result, the UK would only enjoy the first five freedoms of the air.

3 Analysis of Relations Between the European Union Aviation Safety Agency and the United Kingdom.

EASA is a body of the European Union and has legal personality. In each of the Member States, it enjoys the most extensive legal capacity accorded to legal persons under their laws. Furthermore, it may acquire or dispose of movable and immovable property and may be a party to legal proceedings. It should be noted that EASA has its seat in Cologne, Federal Republic of Germany. According to Regulation 2018/1139 (Article 104(4)) [8], the Agency may establish its field offices in Member States or deploy staff in Union Delegations in third countries. As a consequence, EASA has established offices in other countries such as Belgium, Canada, China, Singapore or the United States [9]. The following countries are members of EASA: the 27 EU countries plus Iceland, Liechtenstein, Norway and Switzerland. It should be noted that EASA has contacts with other countries through bilateral agreements, working arrangements, or technical cooperation or a memorandum of understanding. Furthermore, some countries have an observer status with EASA. EASA, within the limits of its competence, assists the Commission in the preparation of draft legislation on civil aviation safety. In addition, it issues opinions and recommendations and also performs its own tasks, e.g. issuing safety and environmental product certifications for aeronautical products (e.g. aircraft, engines) or the approval and oversight of aircraft design and production organisations. However, it should be stressed that the certification of individual aircraft is the responsibility of Member State authorities. It should be noted that the Agency has been granted the power to carry out inspections, training and programmes in the Member States of the European Union. In addition, the EASA also collects data and conducts studies to improve aviation safety. Furthermore, since its establishment, the Agency has been responsible for the certification of aviation equipment and airworthiness. However, it cannot be overlooked that, in accordance with Regulation 216/2008 (now replaced by the Regulation 2018/1139), the Agency has acquired competence in the field of flight crew licensing, air operations and safety assessment of third-country aircraft [10]. Furthermore, the Agency extended its competencies under Regulation (EC) No 1108/2009 of the European Parliament and of the Council of 21 October 2009. As a consequence, issues related to airports, air traffic management and air navigation services have been included in the scope of its competence.

At this stage, it should be stressed that the United Kingdom is no longer a member of the European Union and therefore no longer a member of EASA. According to the withdrawal agreement, it is now officially recognized by the EU as a 'third country' and therefore is no longer participating in the EU decision-making process [11]. However, the EU and the UK have jointly agreed on a transitional period which will last until 31 December 2020. As a result, the question arises of the kind of relationship the UK will have with EASA until and after 31 December 2020. EASA is an agency of the European Union, which aims to ensure the safety of civil air traffic [12]. Until now, EASA has been authorized to issue licences and certificates, but when the United Kingdom leaves the European Union, this situation will change. Currently, the secondary legislation that regulates aviation safety is Regulation (EU) 2019/494 of the European Parliament and of the Council of 25 March 2019 on certain aspects of aviation safety in respect of the withdrawal of the United Kingdom of Great Britain and Northern Ireland from the Union [13]. This act contains provisions on aviation certificates and training modules which will be temporarily respected. It should be noted that under BASA Agreement [14], the U.S. and EU safety agencies, the Federal Aviation Administration (FAA) and the European Aviation Safety Agency (EASA) respectively, accept each other's certification and approval procedures, and only one approval is required. After the transitional period, British airlines will no longer be subject to EASA regulation, and therefore the FAA will no longer be able to rely on inspections by the CAA at FAA-licensed repair stations in the United Kingdom. FAA inspections would be required, as well as CAA approval. As a consequence, the United Kingdom has to take over the responsibilities previously entrusted to EASA in order to re-issue licences. The UK Civil Aviation Authority would thus take over the role of the UK airline regulator. Consequently, the CAA would have to develop new airline legislation. In addition, new staff and additional financial costs will be required for the CAA to operate smoothly. In turn, the EASA can only issue certain certificates on the basis of a third country licence. The extended validity period is therefore intended to reduce the time taken by EASA and operators to issue certain certificates under the new status of the United Kingdom as a third country. The question at this stage is whether the international aviation regulatory authorities will accept the regulatory standards to be established by the CAA. Will the CAA fail to adapt its regulations to those established by EASA? Unfortunately, in this case there is a risk that British airlines may be cut off from flights to certain countries. Given the above considerations, it seems that the best solution would be for the UK to remain in the EASA system as a non-EU country. Therefore, any existing system of mutual recognition between the UK and EASA Member States would remain valid.

4 Analysis of Present Relations Between the United Kingdom and Other Aviation Organisations

The United Kingdom's withdrawal from the European Union does not entail the loss of membership in other aviation organisations such as ICAO (International Civil Aviation Organisation) [15], EUROCONTROL (The European Organisation for the Safety of Air Navigation) [16], or ECAC (European Civil Aviation Conference) [17].

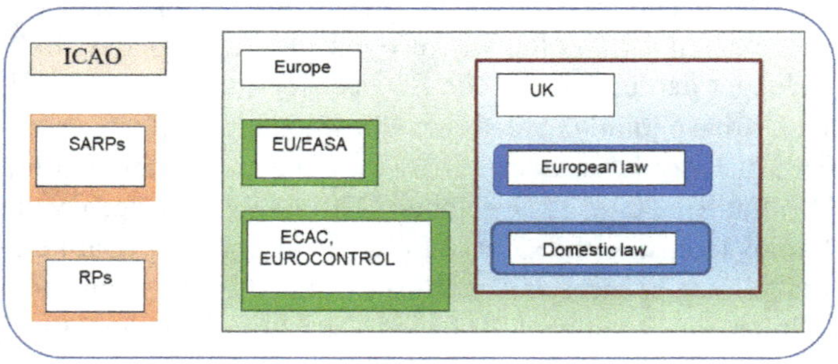

Fig. 1. A scheme of aviation organisations [18].

When analysing the above diagram, it can be seen that the United Kingdom is bound by regulations established by ICAO. In addition, until 1 February 2020, it was a member of EASA and the EU and thus applied the regulations they issued. It should be noted that at European level, the UK continues to be subject to EUROCONTROL and ECAC, of which it is a member (Fig. 1).

4.1 UK Relations with ICAO

ICAO is a UN specialised organisation which develops and implements international regulations governing the safety of international air navigation and supports the development of air transport to ensure safe and orderly development. It should be noted that ICAO issues inter alia Standards and Recommended Practices (SARPs), which are part of the annexes to the Chicago Convention. The legal nature of SARPs has been widely discussed in international doctrine. Most representatives of the doctrine assume that the standards and methods that have their origin directly in the provisions of the Chicago Convention are of a binding nature, and that the Council's power in this respect is a legally binding power. Consequently, it can be concluded that an international organization performs its regulatory function towards its members through the creation of SARPs. Thus, these standards can be qualified as sources of international law. If there are no grounds for an unconditional determination of the nature of the SARPs as binding on ICAO Council Member States, ICAO Member States may not adopt them or implement them in their internal policies. It can therefore be concluded that the abovementioned instruments will serve the purpose of exercising quasi-legislative powers [19]. In accordance with Article 37 of the Chicago Convention, each Contracting State undertakes to cooperate with a view to ensuring the highest practicable level of uniformity of regulations, standards, rules of conduct and organisation with regard to aircraft, personnel, airways and auxiliary services in all cases where such uniformity may facilitate and improve air navigation. It should be noted that there is an exception to this Article. Under Article 38, each State has been granted the power to make changes to regulations or rules of conduct if they cannot be applied in the version recommended by ICAO. It is thus left to the States themselves to implement the international standards and recommended practices in the legal orders of ICAO Member States. It should be noted that the Chicago Convention does not indicate any procedures in this respect. It merely provides guidance for States to comply as far as possible with the rules established from time to time under the Convention. In the case of the United Kingdom, the basic civil aviation legislation is the Civil Aviation Act of 1982. Under that Act, the Secretary of State is the chief authority responsible for the development of civil aviation. The specific tasks contained in the Act are delegated to the Civil Aviation Authority (CAA). In addition, the Act also contains direct obligations under national law to implement the provisions of the Chicago Convention and its relevant Annexes, established in accordance with the applicable procedure. Consequently, the SARPs contained in the Annexes become standards and methods implemented in UK national regulations.

4.2 Relations with EUROCONTROL and ECAC

EUROCONTROL was established by the International Convention on Cooperation for the Safety of Air Navigation, signed in Brussels on 13 December 1960. It is a regional, specialised intergovernmental organisation which aims to coordinate the cooperation of the Member States, as well as international organisations (including the European Union), to ensure the safety of civil and military air navigation. The founding states are: Germany, Belgium, France, the United Kingdom and Luxembourg. EUROCONTROL currently brings together 41 countries to promote close cooperation in air traffic management across the European continent. It should be noted that the UK's membership in EUROCONTROL is independent of its membership in the EU. Once the UK leaves the EU, UK's air navigation services should not be disrupted. It is important that the UK will also remain a full member of EUROCONTROL. The European Civil Aviation Conference, based in Paris, was founded in 1955. It harmonises civil aviation policies and practices among its Member States. Furthermore, its mission is to promote the continuous development of a safe, efficient and sustainable air transport system in Europe.

The United Kingdom has been a member of ECAC since its foundation. Withdrawal of the UK from the Union will not entail the loss of ECAC membership.

5 Flight Safety

Flight safety plays an important role in air transport. Every passenger who goes on a journey wants to get to their destination safely. In order to achieve this, the international community creates numerous regulations to increase the level of safety. In addition, at every stage of the journey, from the first airport, through the flight deck to the destination, the airport operator, the airline, the flight crew, take care of our comfort and safety level. According to Annex 19 to the Chicago Convention [20], safety is a state in which risks associated with aviation activities, related to, or in direct support of the operation of aircraft, are reduced and controlled to an acceptable level. As a consequence of this regulation, one establishes procedures and standards that must be followed by the entire aviation community. Furthermore, the European Parliament and the Council introduced Regulation No 1008/2008 of 24 September 2008 on common rules for the operation of air services in the Community, which governs the licensing of Community air carriers, the right of Community air carriers to operate intra-Community air services and the pricing of intra-Community air services [21]. Consequently, no undertaking may carry passengers, mail or cargo for remuneration or hire unless it has been granted the appropriate operating licence. In this way, the European Union wanted to influence the aspect of air safety because certain conditions have to be met to obtain such an operating licence from the competent authority of a Member State. Thus, the period of validity of such an operating licence lasts as long as all the conditions are duly fulfilled. However, the competent authority monitors this situation in specific cases:

(a) two years after a new licence has been granted;
(b) if a potential problem is suspected; or
(c) at the request of the Commission.

As a consequence of an inspection, the competent licensing authority suspends or revokes the operating licence if a Community air carrier knowingly or negligently provides the competent licensing authority with information of which the relevant details are not correct. Furthermore, it may, for example, suspend or revoke the operating licence of a Community air carrier if it no longer satisfies the requirements of good repute. At this stage, it should be noted that, in line with the opinion of the European Economic and Social Committee (2019/C 190/06) [22], the importance of international aviation for economic growth has been underlined. Furthermore, the European Economic and Social Committee has indicated that civil aviation safety can be ensured by uniform standards implemented by particular countries. The opinion highlights the role of the European Union Aviation Safety Agency, which issues aviation certificates.

6 Conclusions

Air transport is one of the fastest growing methods of transport. Moreover, this type of passenger transport is seen as one of the safest. This is due to the fact that all EU Member

States and third countries apply the standards set by ICAO and EASA. However, when one of the States withdraws from the international organisation, some changes have to occur in the relationship that has existed between them so far. The United Kingdom, which was the first ever to leave the European Union, will face many challenges after the transitional period. First and foremost, it will have to conclude agreements with the countries which concern both air transport and crew training. Another issue will be the regulation of relations with EASA and ECAA. It should be emphasized that EASA not only cooperates with the Member States, but also has contacts with countries under bilateral agreements (Brazil, the United States), working arrangements (Albania, Australia), technical cooperation (Armenia, Burkina Faso) or memorandum of understanding (Honduras, Maldives). Consequently, EASA and the United Kingdom will have to determine the nature of their membership in the Agency and the rules of cooperation. In addition, the United Kingdom will have to establish the conditions for its participation in the ECAA. Becoming a member of the ECAA would allow flights between the signatory states.

Given the air traffic statistics, the United Kingdom cannot afford to break off contacts with the countries that are members of the European Union, as the number of passengers handled is increasing year on year, which makes it necessary to expand the airports and which has an impact on the economy of the whole country. It is reasonable to note that it is only after the end of the transitional period that we will be able to see how the United Kingdom's relations are regulated after 31 December 2020. It will then be clear what legal decisions have been taken to maintain safety standards for flights from and to the UK.

References

1. Treaty on European Union. Maastricht. 1992.02.07., OJ 2004.90.864/30 of 2004.04.30
2. Treaty on the Functioning of the European Union. Rome. 1957.03.25., OJ 2004.90.864/2 of 2004.04.30
3. Regulation (EU) 2019/502 of the European Parliament and of the Council of 25 March 2019 on common rules ensuring basic air connectivity with regard to the withdrawal of the United Kingdom of Great Britain and Northern Ireland from the Union Official Journal of the European Union, L 85 I/49, 27 March 2019
4. Agreement on the withdrawal of the United Kingdom of Great Britain and Northern Ireland from the European Union and the European Atomic Energy Community, OJ EU.L.2020.29.7 of 2020.01.31
5. Political Declaration setting out the framework for the future relationship between the European Union and the United Kingdom (2020/C 34/01), Official Journal of the European Union, C 34/1, 31 January 2020
6. US-European Community. Air Transport Agreement. Washington. 2007.04.30. OJ EU.L.2007.134.4 of 2007.05.25
7. Multilateral Agreement between the European Community and its Member States, the Republic of Albania, Bosnia and Herzegovina, the Republic of Bulgaria, the Republic of Croatia, the former Yugoslav Republic of Macedonia, the Republic of Iceland, the Republic of Montenegro, the Kingdom of Norway, Romania, the Republic of Serbia and the United Nations Interim Administration Mission in Kosovo on the establishment of a European Common Aviation Area, done in Luxembourg on 9 June 2006, OJ 2017.2258. https://sip.lex.pl/akty-prawne/dzu-dziennik-ustaw/wielostronna-umowa-miedzy-wspolnota-europejska-i-jej-panstwami-18658403

8. Regulation (EU) 2018/1139 of the European Parliament and of the Council of 4 July 2018 on common rules in the field of civil aviation and establishing a European Union Aviation Safety Agency, and amending Regulations (EC) No 2111/2005, (EC) No 1008/2008, (EU) No 996/2010, (EU) No 376/2014 and Directives 2014/30/EU and 2014/53/EU of the European Parliament and of the Council and repealing Regulations (EC) No 552/2004 and (EC) No 216/2008 of the European Parliament and of the Council and Council Regulation (EEC) No 3922/91 (Text with EEA relevance), OJ L 212, pp. 1–122, 22 August 2018
9. https://www.easa.europa.eu/easa-and-you/international-cooperation/easa-by-country. Accessed 25 Jul 2020
10. Regulation (EC) No 216/2008 of the European Parliament and of the Council of 20 February 2008 on common rules in the field of civil aviation and establishing a European Aviation Safety Agency, and repealing Council Directive 91/670/EEC, Regulation (EC) No 1592/2002 and Directive 2004/36/EC (Text with EEA relevance). https://eur-lex.europa.eu/legal-content/pl/TXT/?uri=CELEX%3A32008R0216.
11. https://eur-lex.europa.eu/legal-content/PL/TXT/PDF/?uri=CELEX:12019W/TXT(02)&from=PL
12. Regulation (EU) 2018/1139 of the European Parliament and of the Council of 4 July 2018 on common rules in the field of civil aviation and establishing a European Union Aviation Safety Agency, and amending Regulations (EC) No 2111/2005, (EC) No 1008/2008, (EU) No 996/2010, (EU) No 376/2014 and Directives 2014/30/EU and 2014/53/EU of the European Parliament and of the Council
13. Regulation (EU) 2019/494 of the European Parliament and of the Council of 25 March 2019 on certain aspects of aviation safety with regard to the withdrawal of the United Kingdom of Great Britain and Northern Ireland from the Union, Official Journal of the European Union, L 85 I/11, 27 March 2019
14. Agreement between the United States of America and the European Community for cooperation in the regulation of aviation safety. https://eur-lex.europa.eu/legal-content/EN/TXT/PDF/?uri=CELEX:22011A1109(01)&from=EN
15. https://www.icao.int/about-icao/Pages/member-states.aspx
16. https://www.eurocontrol.int/search?keywords=member%20states&sort_by=search_api_relevance
17. www.ecac-ceac.org
18. Diagram based on https://www.icao.int/Meetings/SingaporeSeminar2019/Documents/3_1_Kate_Staples_-_Brexit.pdf
19. Krasicka, I.: Standardy żeglugi powietrznej Organizacji Międzynarodowego Lotnictwa Cywilnego (ICAO), Seria: Monografie Prawnicze, p. 186 (2019)
20. Annex 19 to the Convention on International Civil Aviation
21. Regulation (EC) No 1008/2008 of the European Parliament and of the Council of 24 September 2008 on common rules for the operation of air services in the Community (Recast). Official Journal of the European Union, L 293/3, 31 October 2008
22. Opinion of the European Economic and Social Committee on the 'Proposal for a Regulation of the European Parliament and of the Council on certain aspects of aviation safety with regard to the withdrawal of the United Kingdom of Great Britain and Northern Ireland from the Union' (COM(2018) 894 final - 2018/0434 (COD)) (2019/C 190/06). Official Journal of the European Union, C 190/37, 5 June 2019

Modeling of Ground Handling Processes in Simio Software

Anna Kwasiborska(✉) and Jakub Postół

Faculty of Transport, Warsaw University of Technology, Koszykowa 75 Street, 00-662 Warsaw, Poland

{anna.kwasiborska,jakub.postol.stud}@pw.edu.pl

Abstract. More and more attention is being devoted to increasing the capacity of airspace and airports (airport capacity) due to the increase in air traffic. The effects of the increase in global air traffic are felt not only when managing the movement of aircraft in the air, but also when performing operations at airports. An increasing number of airports must solve the problem of airport capacity. This problem can be solved by developing the airport infrastructure but it is extremely important to analyze processes at the airport. Airports have the necessary infrastructure to enable safe and efficient implementation of airport traffic. Airport traffic is divided into the movement of aircraft performing take-off and landing operations, the movement of aircraft during taxiing after landing and before departure and apron operations. The aircraft handling process is combined with the handling of passengers, luggage, mail and other cargo. Particular attention should be paid to the ground handling process and coordination between relevant entities. Ground handling includes activities performed when an aircraft is parked at an aircraft stand. Among the operations at the aircraft stand, one can distinguish several categories of services performed at the airport for airlines. There is a close relationship between aircraft service at an aircraft stand, taxiway and runway operations. Operations occurring in ground aerodrome traffic can be described as a sequence of operations elementary for aircraft movement in the airport area and on aprons. The organization of ground handling operations performed at an aircraft stand affects the moment of completion of ground handling and the issue of aircraft take off authorization. The identified aerodrome operations at the airport interact with each other to form a structure consisting of a series of activities occurring in parallel or in series. The article examines the process of ground handling of aircraft and includes an analysis based on real-time measurements of steps taken during ground handling. A model was developed and a ground handling model was simulated using Simio software. The developed model was verified on real measurements taken at an airport. An analysis of the duration of performance of individual activities was carried out for different types of aircraft. On this basis, the reliability of the model was confirmed and several simulations were performed. As part of the study of the results, critical paths were analyzed. The model will be used to introduce interference and analysis processes involving a stand to estimate the impact of interference on the timely execution of elementary operations.

Keywords: Simulation software simio · Aircraft ground handling · Critical paths · Functional network

© The Author(s), under exclusive license to Springer Nature Switzerland AG 2021
A. Kwasiborska et al. (Eds.): ATE 2020, LNITI, pp. 57–75, 2021.
https://doi.org/10.1007/978-3-030-70924-2_6

1 Introduction

Every day, several hundred thousand operations are performed at airports around the world. The very rapid increase in the number of flight operations causes problems related to the saturation of current airport infrastructure, which results in delays in flight operations. This applies to both the infrastructure related to passenger service (passenger terminal capacity) and the infrastructure related to ramp service of aircraft and infrastructure directly related to take-off and landing operations. Therefore, one of the key objectives of the aviation industry to improve the efficiency of airports. The most obvious solution is the expansion of airports, e.g. Expanding the passenger terminal, construction of the runway and taxiways or increasing the number of aircraft stands and availability of equipment GSE. However, this is also the most expensive solution.

Another approach to the ever-increasing number of operations at airports is to increase the efficiency and quality of operations at the airport. The authors focused on ground handling of aircraft stands. Ground aerodrome traffic operations are interrelated and the proper performance of each of them affects the punctuality of the others. The moment of completion of ground handling operations determines the start of aircraft taxiing. Then the aircraft begins the take-off operation. It is associated with landing operations of other aircraft. Low efficiency and quality of ground handling services can cause delays. Long delays during ground handling cause deviations from the planned flight schedule. An aircraft usually operates more than one flight a day on medium and short haul routes. This results in accumulation of delays from the departure airport to the arrival airport. During ground handling at the destination airport, delays are increased causing disruption to the organization of the airport. In addition, meteorological disturbances may occur, which may cause additional delays. This situation causes losses for the airline, the ground handler and the airport operator.

The ground handling process is very complicated and consists of many elementary activities. During the ground handling of the aircraft, several different GSE and types of equipment are used. Each of these vehicles should be set in the right place at the right time, properly connected to the aircraft and started. A ground handling agent serves several or over a dozen airlines as well as several types of aircraft. Managing and coordinating all activities during servicing at an aircraft stand [1]. Therefore, proper communication between controllers, airport services and handling agents is also an important issue during ground handling. In order to improve coordination of activities at the airport, appropriate steps are taken to improve operations at the airport. One such solution is the A-CDM solution that is being increasingly implemented, allowing the creation of a common data exchange platform to improve communication between users at the airport. Efficient coordination of services allows synchronization of operations performed at an aircraft stand with operations on the runway [2]. Coordination by providing more information allows for more efficient allocation of airport resources. The impact of intensive air traffic growth on flight delays is noticeable in EUROCONTROL reports. According to a report [3] in 2019, about 46.4% of departures were delayed and there was 40.6% of delayed landings. However, in both cases it is a decrease of approx. 5% compared to the previous year. In 2018, record high delays were recorded and according to the Report [3], actions taken by air carriers contributed to reducing delays in 2019. The average departure delay increased by 1.6 min and in 2018 amounted to 13.1 min. However, the average delay in arrival also decreased by 1.6 min and amounted to 12.2 min. In the report, the reasons for delays are divided into 2 groups: reaction (backward) and primary. According to the Report [3], the main reason for delays is reaction delays (5.7 min on average for flight operations). Reaction delays are the result of the previously mentioned accumulation of delays caused at the departure airport on the way to the arrival airport. The second reason with the greatest influence on general delays are airline delays (3.41 min on average) which are primary delays. In this category, there are delays associated with the handling of passengers and their luggage, ramp service of the aircraft and related to damage to the aircraft and ground infrastructure. However, long delays of over 60 min affected only 5.3% of flights (up from 4% in 2017). The report [3] also indicates on the list of airports with the largest delays in flight operations, e.g. smaller London Stansted and Luton airports, Budapest airport or Portuguese airports in Lisbon and Porto. The article analyzes the ground handling process focusing only on aircraft ramp services. Using the Simio software, a ground handling model has been developed that will form the basis for analyzing interference during ground handling.

1.1 Literature Review

There are many scientific publications concerning the area of airport capacity and performance. The impact of ground handling on airport capacity is increasingly gaining recognition in scientific publications. Many ground handling research and publications focus on selected elements of ground handling. Boarding of passengers is a frequently analyzed process [4]. Boarding may take longer for heavy aircraft than other activities, e.g. refueling or deck equipment. The author of the paper [5] found that a more effective way is to allow passengers to load baggage in several simultaneous storage points along the passenger cabin. Additionally, by pointing out that the traditional way letting passengers in from back to front is not very effective. In contrast, the authors of the publication [6] presented a method that allows determining how much time one should wait for late passengers during boarding. There are also papers related to other activities performed during aircraft ground handling. In the article [7], three ground handling activities were analyzed - boarding, loading baggage and cleaning the deck. The authors proposed a number of changes regarding infrastructure, equipment or work organization of the employees responsible for the above-mentioned activities. They propose, among others, increasing the number of pockets in the vests worn by deck cleaning staff, or applying heijunka rules when scheduling luggage staff [8]. Activities at the end of aircraft ground handling were also considered by scientists. The publication [9] presents a model for optimal planning of pushback of an aircraft from an aircraft stand. The authors have created a mathematical model for the allocation of vehicles to positions. The model also takes into account all operational restrictions. The authors managed to reduce vehicle travel time by 30% and reduce delays while increasing the cost of operation of pushback vehicles.

A significant number of articles have also been written to investigate ground handling services more widely. Such papers do not focus on individual activities, but examine the operation of aircraft ramps as a whole. The authors of [1] analyze the ground handling procedure to check which components are suitable for automation. Automation aims to reduce delays and the costs associated with them, as well as the costs resulting from the employment of personnel. In addition, by reducing the staff at the counter you can reduce the number of disruptions and accidents. The authors argue that ground handling automation must proceed gradually. They suggest that currently, only a part of the activities can be automated, e.g. automatic opening, closing doors, but regulations and certification make it difficult or impossible to implement such a solution. More forward-looking plans include the autonomous movement of ground handling vehicles and automatic connection to aircraft. Paper [4] presents a simulation model of ground handling considered as a single process. The model developed by the authors allows to determine the probability of delays and their causes. The reference airport for the created model was the airport in Wrocław. The same authors also developed another model, this time related to the activities of the handling agent [10]. They used the FlexSim simulation package for this. The model allows optimization of logistics processes related to ground handling. In addition, it enables dynamic GSE assignment to supported aircraft. According to the authors, the model can be used in forecasting the implementation of the connection network over a long time horizon (e.g. next season). However, the created model has some simplifications. The model does not take into account potential interference during the performance of individual operations on the aircraft. The authors assumed a fixed time of performing these activities and did not take into account the reliability of devices used during ground handling. In his work, Justesen [11] dealt with the allocation of ground handling resources at the Copenhagen-Kastrup Airport. Justesen's paper covers a very wide range, including, apart from GSE, check-in, gate and taxiways. The author has developed a complicated algorithm for the allocation of operational resources. It was tested at the Copenhagen Airport and also at the Frankfurt Airport. In both cases, this algorithm increased the resource allocation efficiency by approximately 10%. In contrast, in order to improve communication between aerodrome traffic users and reduce delays, article [12] proposes a decentralized ground handling operations management system based on fuzzy logic. Activities are divided between three users: air traffic controllers, ground vehicle fleet coordinators and the station manager. According to this idea, controllers deal with flight schedules and transfer changes to airport services. Fleet coordinators and station manager communicate with each other to effectively plan and perform all activities. The authors of this article [13] decided to use the comprehensive MATLAB package to develop a ground handling model. The authors developed their own simulation language (SLAGOM) and conducted a number of simulations. The creators of the model emphasize that the developed model can be easily modified and adapted to many airports in the world. In addition, they argue that the model can be used for more efficient planning and prediction of delays. Article [14] deals with the ground handling process through critical path analysis. Critical path analysis showed that the biggest focus should be on operating the passenger compartment to shorten the critical path and thus the service time. The authors present a number of proposals to shorten it. They consider 2 cases - narrow-body (Boeing 737-800) and wide-body (Boeing 747-400) aircraft. The authors' calculations show that it is possible to shorten the service time by 42% for Boeing 737 and by 38% for Boeing 747.

The paper [15] presents a ground handling management model and a graphical interface for ground handling employees and coordinators. The interface allows the manager to monitor the state of the fleet, the allocation of employees, and to send instructions to employees. In addition, the A-CDM concept has been implemented in the work. The use of A-CDM gives better predictability of events and greater operational awareness. In the article [16], the introduction of A-CDM is considered to be an important factor for increasing capacity and coordinating airport operations. The authors point out the advantages of standardizing the implementation of the A-CDM system at European airports.

1.2 Characteristics of Aircraft Ground Handling Processes

Aircraft perform landing operations and then leave the runway via one of the rapid-exit taxiway or taxiways. Then the aircraft participates in airport ground movement. The aircraft moves according to the current airport taxiing scheme and reaches its designated stand. Aircraft operations begin after the aircraft has been positioned. After their completion, readiness for taking off is reported. After obtaining permission, taxiing and take-off operation is performed, which blocks the runway.

The identified aerodrome operations at the airport interact with each other. Particular attention should be paid to ground handling, which is a process consisting of a series of activities occurring in parallel or in series. For further analysis of ground handling, activities carried out at the aircraft stand were included. At airports, many processes are carried out simultaneously, which can be divided [17] into:

- aircraft handling processes,
- aircraft ground handling processes,
- passenger service processes,
- baggage handling and air cargo handling processes,
- operation and maintenance of airport infrastructure elements,
- customs and border control,
- security control,
- prevention of all acts of unlawful interference against civil aviation.

The implementation of the above-mentioned processes allows, above all, to ensure security during all stages of service. In addition, the priority of airport operations is to ensure the regularity and continuity of airport operations. The European Union legal act strictly regarding ground handling is Council Directive 96/67/EC of 15 October 1996 on access to the ground handling market at Community airports [18]. It specifies, among others, entities that can provide ground handling services, business principles, a list of services rendered during ground handling.

The legal bases for ground handling are contained in legal acts such as the Chicago Convention, Annex 14 to ICAO [19]. In addition, the entity providing ground handling, e.g. a handling agent, relies on the IATA Handbooks of the Ground Handling Agent (ground Handling Manual) [11]. In addition, the applicable rules are described in the Airport Operators' Instructions, the operational documentation of the carriers and SLAs, i.e. the agreement between the airline and the handling agent regarding the assurance of a constant level of services. Key performance indicators for ground handling are [1]:

- punctuality,
- minimizing the use of resources (staff, GSE),
- compliance with the rules of A-CDM,
- avoiding damage to equipment and aircraft,
- avoiding accidents involving staff.

The handling agent is supposed to conduct business in a manner that ensures proper functioning and eliminates interference at the airport. In addition, it is obliged to ensure continuity of ground handling services within the scope of the authorization held [10].

The most general definition of ground handling services is all services provided to airport users at airports [18]. Ground handling services may vary in scope of operations depending on the type of aircraft and airline. Each type of aircraft has predetermined activities and estimated times for their execution, as well as the arrangement and connection of GSE. Smaller aircraft generally have shorter service times.

The most general definition of ground handling services is all services provided to airport users at airports [18]. Ground handling services may vary in scope of operations depending on the type of aircraft and airline. Each type of aircraft has predetermined activities and estimated times for their execution, as well as the arrangement and connection of GSE. Smaller aircraft generally have shorter service times.

A schedule is a form of distribution that helps in planning activities over time. The schedule may indicate only the sequence of activities, may be supplemented with the expected duration, it may also contain the required (or calculated) dates (e.g. time) of the beginning and end of the activity. The schedule helps to organize the scope of activities and relationships between them, and also makes it easier to supervise and detect implementation threats early. The graphic form facilitates the use of the schedule. The most popular way of graphically presenting the schedule is the Gantt chart [20].

In smaller aircraft, some operations may be omitted, e.g. provision of stairs. This is due to the construction of the aircraft. Smaller aircraft models may have their own stairs to board the aircraft [10]. An airline can also influence the time of service because it results from the carrier's standards and policy [10]. Entities performing ground handling of an aircraft try to minimize the time of parking at a parking position by performing parallel activities, e.g. refueling and boarding passengers [21]. Parallel execution of actions allows to shorten the critical path of ground handling. Delaying one activity on the critical path results in delaying the entire sequence [22].

In addition, the service time depends on the available human resources and GSE at the airport. Additionally, the impact on the time of ground handling services may depend on weather conditions (temperature, wind, precipitation, fog, sandstorms) [1]. Ground handling services indirectly affect the capacity of the airport's air part. Minimizing downtime allows you to reduce the costs incurred by airlines. Preparatory operations are carried out before operating the aircraft. The cruise number, number of passengers, parking space number and waiting room (gate) number are specified. The next step is preparation for service. Instructions are prepared for unloading aircraft on the basis of messages that inform about potential dangerous loads in a given flight. Documents needed for the crew to complete the next voyage are being completed. Personnel and equipment to be used during aircraft operation will be completed (GSE) [12]. After obtaining information about the landing of the aircraft, the equipment is sent to the given station and a check is made to see if there are any unwanted foreign bodies (FOD), e.g. stones, plastic bags, pieces of luggage. After performing the above-mentioned activities and stopping the aircraft at the station, the apron service of the aircraft begins. The aircraft is secured with stands placed under the front shin of the aircraft and safety cones. An airplane can be secured when: it is completely stopped at the parking stand, the basic engines and the anti-collision lighting are switched off [22]. Then, before substituting devices, a Walk Around is performed to detect possible damage. During ground handling, communication between the aircraft crew and relevant handling agent staff is be maintained. The communication uses concise messages and the principle of confirming the message heard - read back – is observed.

After obtaining permission to start ground handling, GSE substitution begins. Electricity is supplied to the aircraft using power generators (GPUs). The GPU supplies direct or alternating current using an internal combustion engine depending on the size of the aircraft. It is needed to maintain lighting on the aircraft, as well as for air conditioning, and it is used during the startup of engines. Ground handling companies are required to ensure safe entry and disembarkation. The crew and passengers can board the aircraft and disembark using two different solutions. The first is the use of air bridges often called sleeves. These are devices that on one side are attached to the passenger terminal. They allow direct boarding of people from the terminal to the aircraft and disembarkation of people from the aircraft to the terminal. At the same time, when passengers leave the aircraft, checking of the trunk and unloading baggage and cargo can begin. Passenger unloading devices are positioned on the left side of the aircraft. Luggage unloading devices are positioned on the right. After unloading, the baggage hatches are checked for possible damage. In the case of boarding, the procedure is identical. Baggage and goods are delivered to the baggage hold. Bulk loads are secured with a net to prevent them from moving. This method of protection is used in small aircraft. Loads are transported in special ULD containers/pallets in larger aircraft. The use of ULD containers allows to shorten the loading or unloading time of the aircraft [23]. Conveyor belts are used to load/unload bulk loads that are not packed in a container. ULD loaders/unloaders are used to load ULD containers/pallets. Loading baggage and placing passengers on board cannot be arbitrary. The aircraft must be properly loaded and balanced before flight. Proper performance of these activities determines the safe operation of the aircraft [18]. When unloading, also follow the instructions for unloading the aircraft (from luggage and passengers) to prevent the rear of the aircraft from settling on the asphalt by lifting the front part into the air (Tail Tipping).

The catering exchange can be started when passengers disembark from the aircraft. The possibility of simultaneous departure of passengers and exchange of catering applies to most medium and large aircraft, e.g. Boeing 767 or Airbus A320. Smaller regional aircraft (ATR 72, DHC-8) do not have separate doors for loading food on board. In such cases, catering is supplemented after the passengers disembark. When passengers leave the deck, the cabin is cleaned. Depending on the length of the stop and the demand and the agreement with the handling agent, the cabin service may last several minutes and be limited to the necessary minimum, e.g. cleaning the garbage and refilling toilet accessories. If the aircraft has a longer stop, the cabin can be serviced in a more comprehensive manner. Such service then includes e.g. thorough vacuuming and cleaning of seats and floors, replacement of pillows and blankets.

After the flight, the aircraft is serviced. The condition of installations and devices is checked, e.g. the condition of on-board air conditioning. In addition, the contents of the septic tank are emptied. Water is also drained from the tank and the condition of the water system is checked. Then the tank is filled with fresh water from the water service unit. Depending on the need, the aircraft can be refueled during service. Refueling is a dangerous activity when handling aircraft. Therefore, when refueling, the staff must be extremely careful and it is forbidden to use equipment that could ignite the fuel mixture. Refueling can take place after deboarding together with other activities in parallel, e.g. during deck cleaning and loading of catering. It is also possible to refuel with passengers on board to reduce service time. This involves additional conditions. When refueling with passengers on board, assistance must be provided from the airport fire brigade and all escape doors must be open. Tanker cars or a centralized fuel system are used for refueling (and for emptying fuel tanks). The centralized fuel system consists of underground pipeline networks that draw fuel from fuel tanks distant from the parking stand. In such a system, for refueling, the aircraft is connected to the hydrant on the apron. This eliminates the need for tankers and prevents the occurrence of an incident/accident in the form of a tank collision with SP or other ground handling equipment.

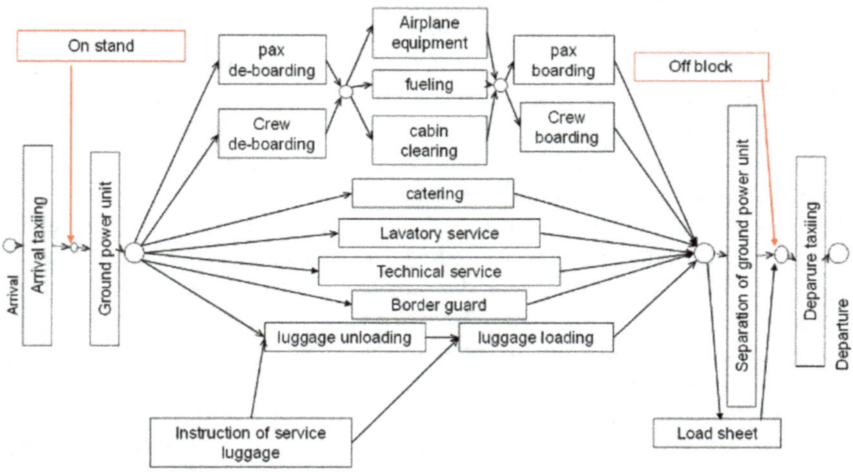

Fig. 1. Scheme of ground handling processes

After completing other activities or during their course, loading of baggage into baggage hatches may begin. Passengers and crew can start boarding after the cabin is cleaned. After loading, a Loadsheet is issued and given to the crew along with the passenger list and other documents. After these documents are provided, the aircraft is ready to start the engine. Figure 1 is a general functional network diagram containing all recorded aircraft ground handling activities.

Skillful planning of start and end times of individual elementary activities affects the end time of all ground handling. The knowledge of the time of termination of ground handling of aircraft has an impact on making correct decisions by airport traffic controllers. The time spent on ground handling depends on the number of passenger seats on the aircraft and the type of voyage. Examples of functional networks of elementary ground handling operations of various types of aircraft are given below. The configuration of the activities will be different depending on the type of aircraft and type of ground handling expected.

When starting the engine, ground crews communicate with the crew through a head set or with the help of their hands. The captain decides which engine will start first. A special ASU vehicle is used to start the engine if the aircraft does not have an auxiliary power unit (APU). An aircraft may begin to pushback when:

- all ground handling activities have been completed,
- all luggage hatches and doors have been closed,
- GSE is disconnected and removed,
- wheels are locked,
- the driver of the pushback tractor has received a signal enabling the pushback.

Some aircraft can leave the stand using their own engines and thrust reversals (Powerback). This is a dangerous activity due to the possibility of pulling the FOD into the engines or pushing them towards ground handling personnel or towards passengers at the terminal [24].

In the winter, from October to March, an additional activity is carried out before the start. During this period, planes are de-iced when they are already covered in ice or there is a risk of icing. Deicing operations can be carried out on request outside of the winter, when the need arises.

Ground handling is a complex process consisting of many activities occurring in parallel or in series. The service coordinator oversees the ground handling process. This is designed to coordinate all activities so they are performed in accordance with the provisions and agreements concluded between users and to comply with the timeliness of ground handling. In addition, it deals with communication with other participants and aims to anticipate and respond to disturbances and threats.

A huge fleet of ground vehicles performing specific tasks is associated with ground handling SP. It is up to the GSE fleet manager to ensure that vehicles are delivered on time to the correct stand, and to avoid and resolve conflicts. In addition, using e.g. the VMS computer support system and GPS sensors, they can track the speed and position of each vehicle and redirect the operator to a different position on an ongoing basis when the need arises. Systems are also used to manage and plan (long-term and real-time) ground staff work schedules [58].

2 Simulation Model of Aircraft Ground Service

2.1 Model Description

The model was made in four versions by checking the effect of the different assumptions. To support the ground handling model, computer support was used in the form of the

Simio package. The critical path for ground handling processes of aircraft belonging to low-cost carriers designated in article A, B, C and D was analyzed. Measurements were made for approx. 80 different flights and concern 4 different narrow-body aircraft: Boeing 737-300 and 737-700 and Airbus A319 and A320. These are medium-sized aircraft taking between 140 and 180 passengers depending on the configuration. The above aircraft are characterized by the fact that they have separate doors for catering, enabling unambiguous entry and provision of catering, which was also included in the models. The passengers' and crew's entry/exit operations are carried out by a stepped staircase and not by a passenger bridge.

Using Simio, a database was created containing 8 to 10 tables depending on the model and its variant, containing the times of elementary ground handling activities. A random flight schedule was prepared taking into account the curfew from 23:30 to 05:30. In addition, the time intervals in this distribution have been selected to minimize the likelihood of simultaneous operation of two SPs. All tables were connected by relation keys, and then the appropriate references to the tables were introduced to the models and their variants described below.

The model made in Simio reflects the activities performed when the aircraft is stationary. It covers almost all services performed during aircraft maintenance. The model does not include operations related to de-icing of the aircraft. This is dictated by the fact that deicing is a seasonal activity, and it is also often carried out at another specially adapted parking stand. Figure 2 shows a fragment of the ground handling model.

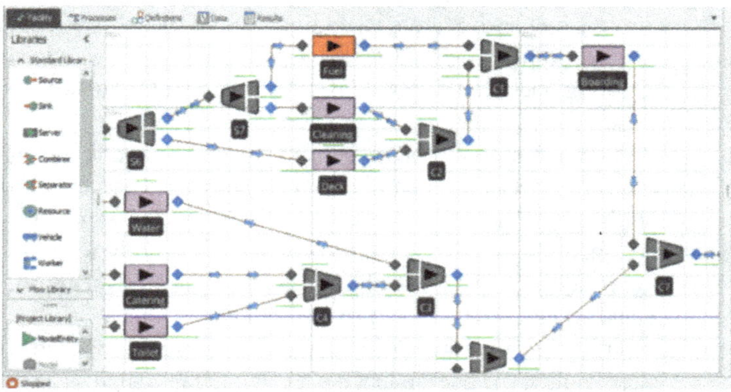

Fig. 2. Part of ground handling model in Simio

The created model uses one source (Source1) and one outlet (Sink1). Units can be generated on a schedule or at fixed or random intervals. The outlet, on the other hand, is used to remove units that have completed a specific sequence of activities in the model. The unit in the model moves specific paths between nodes belonging to Server type objects. The time of performing individual activities is determined in accordance with the relevant tables from the database. Each Server type object is a service station with an input and output buffer, whose capacities can be modified. The model includes refueling without LSP assistance, i.e. refueling can be started when passengers are not on board the aircraft and refueling with LSP assistance enabling simultaneous entry/exit of passengers (Fig. 3). Depending on the type of generated unit, one of the inactive servers will have a set activity time of 0.

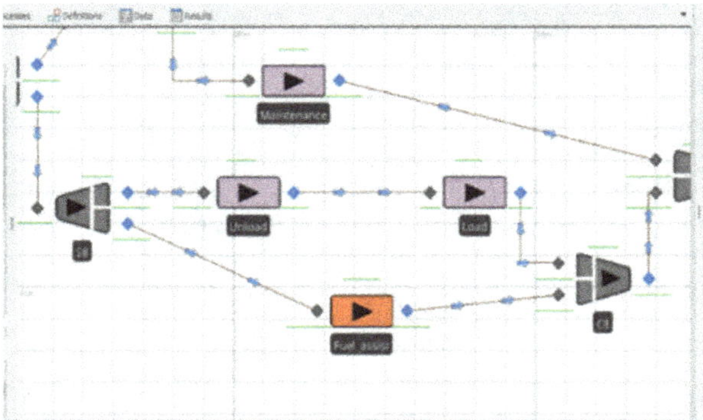

Fig. 3. Scheme of a model with fuel activity

Simulations in the Simio environment began with model verification. As part of this simulation, the average times of apron activities and the average service time of the entire aircraft were compared. Critical paths for each ground handling process simulated in Simio were also investigated. A simulation scenario was created for the adopted variants, assuming five iterations of the flight schedule. For the random variant, the arrival times of units are generated by the pseudo-random number generator in Simio. This time is generated according to the normal distribution. IAMeanTime and STDev parameters (mean and standard deviation) are simulation variables (Controls) in this variant. They can be conveniently changed from the simulation window without having to modify them in the database. Four different simulation scenarios were assumed for the random variant, differing in the parameters of the distribution of Entity units generation time.

2.2 Input Data

The actual data of the operations performed, which included the times of elementary activities during ground handling, was used to conduct the simulation in Simio. These times were obtained by measuring the start and end times of activities during the actual ground handling of aircraft at the airport. The measurements mainly concerned the operation of servicing low-cost airlines. In total, the sample contained approx. 80 measurements for four different narrow-body aircraft. The developed database included the duration of the operations: chocking the aircraft, deboarding, unloading and loading the aircraft, fueling, water and toilet service, catering, and aircraft service. Figure 4 shows the percentage of execution times for selected ground handling operations.

The average times during ground handling operations are shown in Table 1.

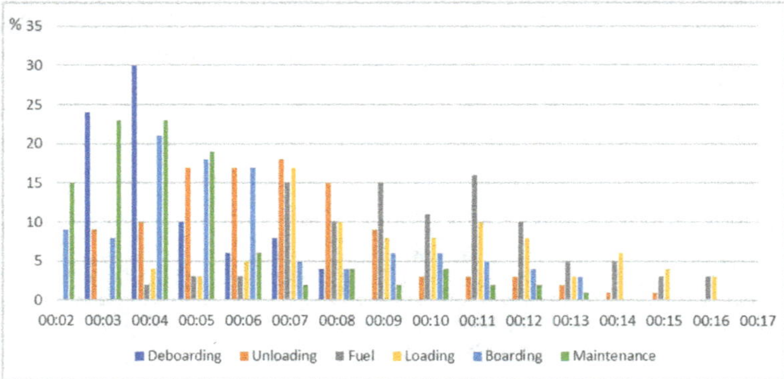

Fig. 4. Histogram of selected ground handling times

Table 1. Summarized the results of actual data with the results of Simio

Elementary activities	Average time [min:s]
Deboarding	04:16
Unloading	07:00
Fuelling	10:04
Maintenance	04:30
Catering	05:11
Toilet servicing	07:02
Water servicing	06:28
Aircraft cleaning	07:02
Aircraft servicing	03:51
Boarding	05:40
Loading	10:30

In Simio, it is possible to build a large database on the basis of data contained in a spreadsheet or in a database in SQL or Microsoft Access. Simio allows to create tables with data and schedules and create relationships between them using appropriate coding.

2.3 Simulation Results

Using Simio software, simulations were carried out for various scenarios. Figure 5 shows a fragment of the results obtained after the simulation in Simio. The columns under the heading Responses show the average duration of each activity after five iterations of the simulation. For the purposes of simulating the model, the Responses columns apart from the duration of individual activities also include the minimum, maximum and average aircraft service time.

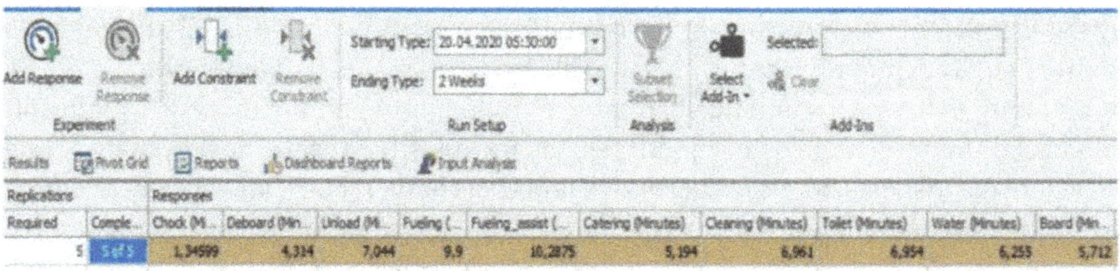

Fig. 5. Fragment of results after the performed simulation

Simio results have been compared with actual measurements in Table 2.

Table 2. Summarized the results of actual data with the results of Simio

	Avarage time	Avarage time Simio	Airbus	Airbus simio	Boeing	Boeing simio
Chocking	00:28,5	01:20,8	00:20,0	00:22,7	00:35,7	00:35,8
Deboarding	04:16,2	04:18,8	04:16,7	04:17,6	04:15,7	04:17,3
Unloading	07:00,0	07:02,6	07:03,3	06:43,6	06:57,1	07:01,9
Fuelling	10:04,8	10:05,6	10:02,7	10:02,1	10:06,0	10:00,1
Maintenance	04:30,0	04:22,3	04:23,3	04:23,9	04:33,5	04:33,4
Catering	05:11,4	05:12,0	05:00,0	05:04,7	05:20,0	05:11,5
Toilet servicing	07:02,9	06:58,2	07:13,8	07:19,4	06:45,0	06:40,6
Water servicing	06:15,5	06:15,3	06:05,1	06:07,4	06:27,6	06:25,2
Aircraft cleaning	07:02,7	06:57,7	07:48,0	07:32,3	06:21,8	06:22,6
Aircraft servicing	03:51,3	03:59,2	03:54,3	03:54,8	03:49,6	03:54,4
Boarding	05:40,0	05:42,7	05:40,0	05:28,2	05:40,0	05:48,2
Loading	10:30,8	10:26,9	10:26,7	10:16,7	10:34,3	10:36,0
Max time	66:00,0	79:24,0	56:00,0	54:24,0	66:00,0	76:12,0
Min time	24:00,0	22:48,0	24:00,0	21:48,0	27:00,0	23:48,0

Differences in performance times are small, reaching a few seconds in most cases. For the average and minimum service time of the SP, the difference between the measurements and the simulation is about 2–3 min. The largest deviations can be seen at the maximum SP service time, where the difference between the simulation value and the actual value is over 13 min. The times of individual activities are close to actual values for all scenarios. The only exception is the duration of the activity. There is a large discrepancy in the values obtained for this activity between individual scenarios. This is dictated by the fact that the waiting time in the buffer for the service is additionally calculated in Simio. Due to extremely short time intervals between successive units, some units had to wait until the previous unit finished service. Extremely short 60 intervals significantly influenced the service times of the SP. This is most visible for the next scenario (Table 3), where the distance between successive units was only 45min. Equations are centered and set on a separate line.

Table 3. Results

Time of handling	Airline A	Airline A Simio	Airline B	Airline B Simio	Airline C	Airline C Simio	Airline D	Airline D Simio
Average	33:54,5	40:33,1	29:00,0	34:21,4	42:51,0	52:31,3	40:54,0	49:15,9
Max	41:00,0	81,42,1	36:00,0	77:12,0	66:00,0	102:22,6	56:00,0	91:36,0
Min	27:00,0	22:12,0	24:00,0	20:00,0	30:00,0	27:12,0	29:00,0	28:24,0

As a result, the C and D airplanes, which are usually operated a little longer, did not finish their service before the arrival of the next aircraft. In the third scenario, the average aircraft service time increased by approx. 7 min relative to the measurement values. However, the maximum service time increased by almost 40 min. In scenarios 1 and 2, the same average value was set, only the standard deviation was changed. The difference of 6 min of the standard deviation reduced the waiting time for service by 2 min. In the case of overall service times in the second scenario, the average service time decreased by about 1 min. And the maximum service time is up to 10 min. In contrast, scenario 4 provides sufficient space (2h) between subsequent Entities. All aircraft managed to end service before the arrival of a new aircraft, which is why the results for the last scenario are similar to the results from the simulations of previous variants. Next, the distribution of critical paths for individual variants was compared. Three different types of critical paths were considered. The column collectively refers to the cumulative results of all options. Comparison with real measurements is presented in Table 4.

Table 4. Critical paths for scenario of model

Processes	Data	Basic	AC	Airline	Random
Deboarding-Aircraft servicing-boarding	96%	68%	69%	70%	76%
Unloading-loading	9%	36%	34%	36%	28%
Fuelling-maintanance	5%	17%	13%	11%	12%

The analysis of critical paths showed that the most common critical path is the path, which is characterized by a sequence of activities related to passenger service, deck and/or refueling (exit path - cabin service/refueling - entrance). Simulation of individual variants of the first model showed very similar results in terms of critical paths. It is a critical path for approx. 70% of measurements regardless of the variant. The path for unloading and loading occurs for approx. 33% of measurements. The generalized path for activities that can be started regardless of the departure of passengers was a critical path for only about 10% of the results.

3 Discussion

The obtained simulation results of the model positively verified the built ground handling model. Making the model in several variants allowed for a broader analysis of results. The variants allowed to check how ground handling times differ depending on the model of the aircraft served or the air carrier. For all variants, the results turned out to be very close to actual measurements. The largest deviations could be recorded for the maximum values of aircraft service time resulting from the randomness of the simulation. In particular, noticeable for carriers A and C. For further verification, a chi-square test was carried out to check compliance of the distributions between the measurement results and the simulation results. A borderline significance level (p-value) of 0.906 was achieved. It is greater than the assumed significance level $\alpha = 0.05$. This means that there is no reason to reject the hypothesis that there is no statistically significant difference between the results for the 95% confidence level.

Based on the results of the simulation of the random variant, it was observed that the shortening of the time between successive aircraft resulted in a significant increase in the average and maximum service time of the aircraft. The change in these values depended on the simulation scenario. In the second scenario, the one hour interval between flights with a small standard deviation was sufficient for most supported aircraft. The average service time increased by about 1 min compared to the results of the fourth scenario. In contrast, the third scenario shows that planes in extreme cases reached service time almost twice as long. However, the increase in these values is due to the specifics of performing simulations in Simio. The service time for a given unit was calculated from the moment it was generated by the source. In this way, the SP service time, which in addition to the times of performing operations, connecting and disconnecting GSE, also includes the waiting time for service, e.g. near this stand.

The analysis of critical paths showed that the most common critical path is the path which is characterized by a sequence of activities related to passenger service, deck and/or refueling (exit path - cabin service/refueling - entrance). Simulation of individual variants of the first model showed very similar results in terms of critical paths. It is a critical path for approx. 70% of measurements regardless of the variant. The path for unloading and loading occurs for approx. 33% of measurements. The generalized path for activities that can be started regardless of the exit of passengers was a critical path only for about 10% of the results. Differences at times of individual paths were varied, in some cases a difference of over 10 min was achieved. On the other hand, many identical track times have been noted. This also results in the value of 100% being exceeded when adding critical path values.

In the article [1], tests of wide-body and narrow-body aircraft were carried out and the results indicated that critical paths are distributed in a completely different way. The dominant critical path is the path related to the unloading and loading of the aircraft. This is due to the fact that wide-body aircraft have much larger baggage hatches and generally higher payload. Airlines use this fact to transport cargo, not limited to passengers' baggage only. Narrow-hull aircraft will not load a significant amount of additional cargo due to their smaller dimensions. Ground handling agents and airlines, especially low-cost carriers, strive to minimize critical paths. A visible example of a shortened critical path is the removal of refueling from the passenger exit sequence - cabin service and/or refueling - passenger entry. Refueling can be removed from this chain by providing LSP assistance during refueling. During the analysis, cases of service with refueling without assistance, with assistance and without refueling were considered. The results of the average service time for units such as Airbus_assist are lower than for units of the Airbus type. Simulation results show that refueling with assistance allows to "save" several valuable minutes.

The first type means ground handling, during which refueling is performed without assistance, i.e. after the passengers disembark from the aircraft. SP_assist means a flight with LSP assistance provided during refueling. In contrast, SP_NF identifies a flight that does not require refueling. In this case, refueling is skipped, other operations are carried out normally.

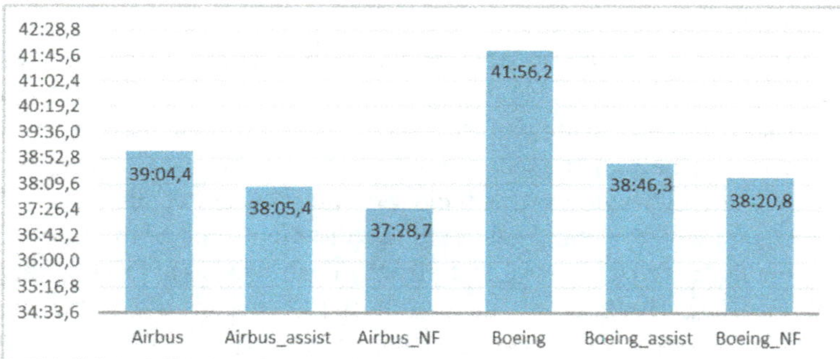

Fig. 6. Aircraft service time depending on the type of service

Figure 6 also shows that non-refueling aircraft are served even faster. This shows that refueling is an important aspect that may affect the total service time of SP. However, the impact of refueling assistance or non-refueling assistance is reduced by the need to clean the deck after passengers disembark. According to Fig. 6, the use of fire brigade assistance allows to reduce the time of SP operation by about 1–3 min. Each airline has its own company policy. Therefore, the operation of the same aircraft model may be different. It also affects the time of aircraft service, which may vary between airlines. Figure 7 shows the average SP service time for various airlines using simulation results.

The type of aircraft served is also an important issue. Each model has times of performing individual activities estimated by the manufacturer. The obvious thing is that larger wide-body aircraft require more time to handle such a model. Using Fig. 8, it was found that the differences are also noticeable in similar-sized airplanes, e.g. between Airbus A320 and Boeing 737 aircraft. The differences in this case are not as noticeable as when comparing Boeing 737 and e.g. a much larger Airbus A380.

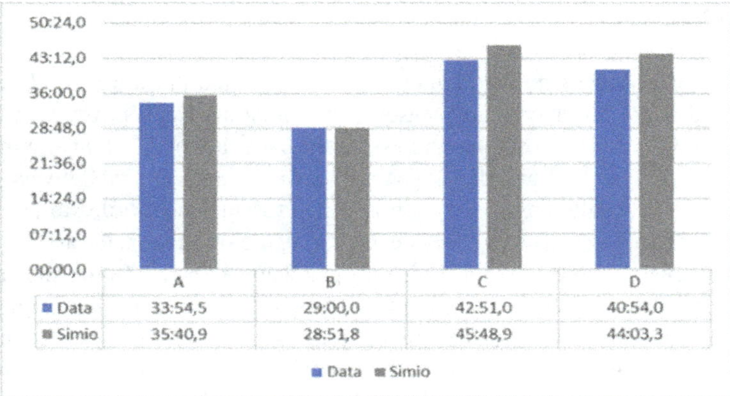

Fig. 7. Average aircraft service time by airline

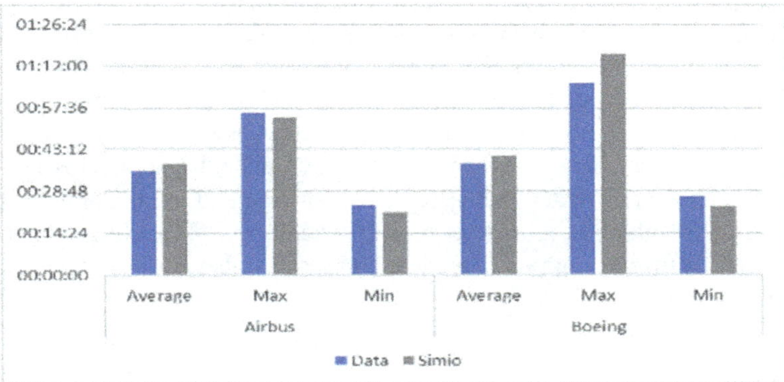

Fig. 8. Average aircraft service time by aircraft

Activities performed during ground handling have not changed for many years. However, the technology used has changed, but the most changed approach in terms of planning and coordinating activities pertains to ground handling SP and other activities at the airport. Due to the multitude of activities and the participation of entities often having conflicting goals and dependence on other services at the airport, ground handling is exposed to interference. Reports show that ground handling is one of the main causes of flight delays. Due to the specificity of transport, delays accumulate between successive take-off and landing operations. Coordination of individual activities with entities involved in ground handling services and with other services at the airport allows to increase the efficiency and quality of ground handling services and other operations and has an impact on airport capacity. The developed model will be extended and it will be possible to analyze ground handling processes in the event of interference.

Please note that the first paragraph of a section or subsection is not indented. The first paragraphs that follows a table, figure, equation etc. does not have an indent, either.

4 Conclusions

In order to perform the simulation, a database was created in Simio based on external data. These data are based on actual measurements of the start and end times of individual ground handling activities for different low-cost airlines and for different aircraft models. A ground handling model was developed with four options. The model was simulated and the obtained results compared with real measurements in order to ascertain the correctness of building this model. Before the start of the simulation, the assumptions and simulation scenarios for the model and the variant were defined in Simio separately. The simulation of the model showed that the simulation results obtained were highly similar to those obtained from actual measurements. The times of activities usually differed by a maximum of several seconds. Larger deviations were observed for extreme values. This confirmed the credibility of the model. As part of the study, critical paths were also analyzed. It was observed that for the considered air carriers, the most frequently occurring critical sequence of operations was the exit-cleaning/refueling-entry sequence. It constitutes a significant majority occurring for about 70% of all results.

In summary, ground handling is a complex process that involves performing a number of different activities simultaneously. All this means that ground handling is exposed to a variety of disturbances that may translate into delays during departure or arrival. Delays related to ground handling are one of the main causes of flight delays in general. Ensuring coordination between personnel, dispatchers and other services is an important element of cooperation. Better exchange of information and joint actions, e.g. via the common A-CDM platform, allows for more efficient ground handling. This translates into more efficient work of ground handling entities, but also into increased capacity of the entire airport.

References

1. Alonso Tabares, D., Mora-Camino, F.: Aircraft ground handling: analysis for automation. In: 17th AIAA Aviation Technology, Integration and Operations Conference, USA (2010)
2. Kwasiborska, A.: Modelling of ground handling operations at airport, Poland. J. KONES **17**(3), 253–261 (2010)
3. Hesselink, H.H., Raoul, J.L., Bowen, D.: A-CDM Standard to Guide Implementation in Europe. National Aerospace Laboratory NLR, The Hague (2017)
4. Kierzkowski, A., Kisiel, T.: A simulation model of aircraft ground handling: case study of the Wroclaw airport terminal. In: Advances in Intelligent Systems and Computing, Wrocław, pp. 109–125 (2017)
5. Steffen, J.: Optimal boarding method for airline passengers. J. Air Transp. Manage. **14**(3), 146–150 (2015)
6. Skorupski, J., Wierzbińska, M.: A method to evaluate the time of waiting for a late passenger. J. Air Transp. Manage. **47**, 79–89 (2015)
7. Nugroho, I., Riastuti, U., Iridiastadi, H.: Performance improvement suggestions for ground handling using lean solutions. Procedia Soc. Behav. Sci. **65**, 462–467 (2012)
8. Norin, A., Granberg, T., Yuan, D., Värbrand, P.: Airport logistics—a case study of the turn-around process. J. Air Transp. Manage. **20**, 31–34 (2012)
9. Du, J., Brunner, J., Kolish, R.: Planning towing processes at airports more efficiently. Transp. Res. Part E Logist. Transp. Rev. **70**, 293–304 (2014)

10. Kierzkowski, A., Kisiel, T.: Model funkcjonowania agenta obsługi naziemnej uwzględniający różnorodności obsługiwanych statków powietrznych . Logistyka **3**, 2900–2909 (2014)
11. Justesen, T.F.: Allocation of Ground Handling Resources at Copenhagen Airport, DTU Management Engineering (2004)
12. Fitouri-Trabelsi, S., Cosenza, C., El Moudani W., Mora-Camino F.: Managing Uncertainty at Airports Ground Handling. Airports in Urban Networks, Service Technique de l'Aviation Civile (STAC), Paris (2004)
13. Voulgarellis, P., Christodoulou, M., Boutalis, Y.: A MATLAB based simulation language for aircraft ground handling operations at hub airports (SLAGOM). In: Mediterranean Conference on Control and Automation, pp. 334–339 (2005)
14. Beelaerts van Blokland, W.W.A., Huijser, R., Stahls, R., Santema, S.A.: Future airport turnaround ground handling processes. In: Conference proceedings of 10th TRAIL Congress, TRAIL Research School, Delft (2008)
15. Alonso, B.: Ground handling management modelling and visual interface conceptual design (2007). https://bibing.us.es/proyectos/abreproy/60213/fichero/PFC+Beatriz+Arias+Alonso+-+Ground+Handling+(ingl%C3%A9s).pdf
16. Hesselink, H.H., Raoul, J.L., Bowen, D.: A CDM Standard to Guide Implementation in Europe. National Aerospace Laboratory NLR, The Hague (2007)
17. Skorupski, J., Stelmach, A.: Deklarowana przepustowość portu lotniczego – problemy, koncepcje rozwiązań, Infrastruktura Transportu, nr 1 (2008)
18. Council Directive 96/67/EC of 15 October 1996 on access to the ground handling market at Community airports
19. Kwasiborska, A., Malarski, M.: Analiza operacji obsługi naziemnej dla zadania koordynacji ruchu lotniskowego, w: Prace Naukowe Politechniki Warszawskiej. Transport, nr 71 (2009)
20. Airbus, A350—Aircraft Characteristics Airport and Maintenance Planning AC; Airbus: Blagnac, France (2017)
21. Kwasiborska, A.: Podstawowe wymogi dotyczące obsługi samolotu Airbus A380, Logistyka, nr 2 (2010)
22. Kwasiborska, A.: Modelling of ground handling operations at airport. J. KONES Powertrain Transp. **17**(3) (2010)
23. IATA AHM 810 Standard Ground Handling Agreement (2013)
24. Transport Research Board of the National Academies: Airport Passenger Terminal Planning and Design: Guidebook. ACRP report 25. Airport passenger terminal planning and design, Guidebook, vol. 1, p. 128 (2010)

Safety First or Safety Sometimes?

Dominika Marzec(✉)

'Polish Airports' State Enterprise, 1 Żwirki i Wigury Street, 00-906 Warsaw, Poland
maarzecdominika@gmail.com

Abstract. The aim of this study was to diagnose the main problems related with safety management system implementation in civil aviation. As it has already been seven years since ICAO Annex 19 was published, the Author analyses performance and usefulness of chosen elements of the SMS. The author conducted the research by participant observation, analysis and expert interviews. Also the surveys concerning performance of specific SMS processes were carried out.

Results show shortcomings of the safety management system related with some of its twelve elements. The presence of problems related to the first SMS component (Safety policy and objectives) is the most alarming.

There are significant differences of the development stage of SMS among the organizations. It is essential that each organization tailors the implementation of safety management to fit their specific environment. Top management of aviation organizations does not always notice the potential benefits of the SMS, which may cause the seclusion of this system. The Safety Management Team should be aware of operational issues and should be part of risk-based decision-making related to the organization's daily business. It is crucial not to treat the SMS as a stand-alone department or an undue burden.

Keywords: Safety management system · Aviation safety · Air transportation

1 Introduction

According to international law, the State of Poland is obliged to implement a safety management system for aviation. There are also many requirements in this scope for organizations that operate within this industry sector. The priority of operation for all aviation entities is given by the law: safety. Aviation is considered to be the safest mean of transport, and maintaining such a state in the light of the ever-increasing size of this mode of transport and the continuously emerging safety hazards is a big challenge for the management. To maintain the position of the safest mean of transporting people and cargo, it is necessary to improve the implemented processes and to use new, more effective tools. A SMS is needed in organizations not only to comply with the law. Taking a broad holistic view, safety management ensures development of aviation businesses. Thanks to the low rate of accidents, passengers trust it and are willing to use it. Once aircraft start distressing the customers, they will travel less using the services of airlines. Consequently different aviation organizations will experience a decline. This makes

© The Author(s), under exclusive license to Springer Nature Switzerland AG 2021
A. Kwasiborska et al. (Eds.): ATE 2020, LNITI, pp. 76–85, 2021.
https://doi.org/10.1007/978-3-030-70924-2_7

businessmen lobby for a change in the law to eliminate entities not assuring a high level of safety.

Research was conducted with the aim to evaluate the performance and usefulness of the SMS in the Polish civil aviation organization. The selection of the research topic was dictated by the need to summarize the most frequent problems related to the SMS functioning in Polish civil aviation organizations in order to point at opportunities for further improvement.

The Author is aware of the broadness of the topic. This article concentrates on the opinions of representatives of chosen aviation organizations of different sizes. The research took into account the experience of Mr. Waldemar Królikowski - a witness and a participant of the introduction of international law, including SMS requirements, into Polish aviation. To diagnose the main problems related with safety management system implementation in civil aviation in Poland, the Author conducted participant observation and analysis as well.

2 Extent of the Field to Be Managed

As the air traffic is continuously growing and the number of potential hazards for safety is also rising, the improvement of management systems is required to ensure an adequate level of safety. According to the statistics of scheduled air traffic published by the International Civil Aviation Organization [7], aircraft took off 37 823 000 times in 2018 (23,6% of them in Europe). In total, they carried 4 322 354 000 passengers (26% of them in Europe).

The European Union Aviation Safety Agency specified in the Annual Safety Review 2019 [3] the top key risk areas which are the most frequent sources of accidents or potential sources of accidents. The following Table 1 gives a summary of identified key risk areas by domain in descending order.

As we can see, the risk catalogue is wide and every aviation domain needs an SMS adapted to its specificity. On the other hand, we need to remember that safety of one domain depends on another domain's activity.

There are also new challenges for aviation identified by EASA in the last Annual Activity Report [2]:

- introduction of drones,
- cybersecurity risks,
- digitalization.

It may be noticed that not all hazards and challenges are easily foreseeable. For instance: who expected such a drop in operations in 2020 caused by the COVID-19 epidemic of? However, we are able to predict an increase of the probability of safety related occurrences, the cause of which will be the human factor in the form of an unintentional error. Stress, job insecurity and decline in air traffic are contributing factors. We already observed such a phenomenon related to the terrorist attacks of September 11 2001, the Gulf War and the SARS epidemic in 2003, the financial crisis of 2008 and the eruption of the Eyjafjallajökull volcano in 2010 [9]. Another example could be the

Table 1. Summary of key risk areas by domain.

Aero planes	CAT (Airline, Air-Taxi and NCC Business)	Specialized Operations	Non-commercially operated small aero planes
higher risk ↑	Aircraft Upset	Aircraft Upset	Aircraft Upset
	Runway Excursion	Airborne Collision	Terrain Collision
	Security	Terrain Collision	Obstacle Collision in Flight
	Runway Collision	Obstacle Collision in Flight	Runway Excursion
	Airborne Collision	Runway Excursion	Airborne Collision
	Injuries/Damages	Ground Collision	Aircraft Environment
	Aircraft Environment	Aircraft Environment	Ground Collision
	Ground Collision	Taxiway Excursion	Taxiway Excursion
↓	Taxiway Excursion	Runway Collision	Runway Collision
	Terrain Collision		
lower risk	Obstacle Collision		

Helicopters	Offshore commercial air transport	Commercial operations - other than offshore	Specialized Operations	Non-commercial operations
	Aircraft Upset	Terrain Collision	Aircraft Upset	Aircraft Upset
higher risk ↑	Excursions	Aircraft Upset	Terrain Collision	Terrain Collision
	Obstacle Collision in Flight	Obstacle Collision in Flight	Obstacle Collision in Flight	Injuries/ Damage
	Injuries/ Damage	Airborne Collision	Excursions	Obstacle Collision in Flight
	Ground Collision	Excursions	Injuries/ Damage	Excursions
	Terrain Collision	Injuries/ Damage	Ground Collision	Ground Collision
↓	Airborne Collision	Aircraft Environment	Airborne Collision	Airborne Collision
lower risk	Runway Collision	Ground Collision	Runway Collision	Runway Collision
	Aircraft Environment	Runway Collision	Aircraft Environment	Aircraft Environment

	Balloons	**Sailplanes**	**Aerodrome and Ground Handling**	**ATM/ANS**
higher risk ↑	Balloon Landings	Aircraft Upset	Ground Collision	Runway Collision
	Obstacle Collision in Flight	Landing Area Excursions	Aircraft Upset	Airborne Collision
	Balloon Upset	Terrain Collision	Runway Excursion	Runway Excursion
	Terrain Collision	Obstacle Collision in Flight	Obstacle Collision in Flight	Terrain Collision
↓	Balloon Environment	Airborne Collision	Terrain Collision	Injuries/ Damage
lower risk	Airborne Collision	Aircraft Environment	Aircraft Environment	Obstacle Collision in Flight
			Airborne Collision	Ground Collision
			Taxiway Excursion	Aircraft Upset
			Runway Collision	Taxiway Excursion

sudden loss of consumer confidence related to Boeing 737 MAX accidents in 2018 and 2019.

To improve our systems, we first need to be sure they are as effective as they are today. We cannot assume that processes established seven years ago (or a year ago) are appropriate to manage aviation safety today.

3 Difficulties of SMS Implementation

In 2019, the USOAP (Universal Safety Oversight Audit Programme) celebrated its twentieth anniversary. USOAP is an ICAO programme, which is considered as the first global initiative to improve aviation safety. The programme consisted in regular and mandatory audits of ICAO Member States' safety oversight systems. The impact of this programme on aviation safety is tremendous, even though ICAO does not penalize states. Reports of the audits are published – aviation entities monitor them, and they take into account ICAO audit results when deciding about the expansion of their businesses. It is a sufficient argument for the states to seek to obtain the highest results [10].

The first audit under USOAP programme in Poland was performed by ICAO in 2000. The results unveiled obsolete aviation regulations (the ones introduced in 1965 were still in force!) which did not meet the basic standards of international aviation. The shortages in the Polish legal system presented a significant risk for the safety of aviation. The ministry responsible for aviation was experiencing many structural and personal changes at that time, which seemed to prevent improvement of aviation regulations. Only in 2002 «The Act on Aviation Law» became an integral part of Polish legal system. This regulation was the starting point of the repair plan prepared in response to results of the USOAP audit. A follow-up audit was performed in 2003. It confirmed the progress of aviation law and refinement of the oversights of Polish aviation [11].

The first legal requirements for the safety management system were introduced when the ICAO implemented it (2006) in the framework of Annexes to the Convention on International Aviation:

- Annex 1 - Personnel Licensing;
- Annex 6 - Operation of Aircraft;
- Annex 8 - Airworthiness of Aircraft;
- Annex 11 - Air Traffic Services;
- Annex 13 - Aircraft Accident and Incident Investigation; and
- Annex 14 - Aerodromes.

The ICAO High-level Safety Conference 2010 proposed the development of a new Annex (19) dedicated to safety management [1]. The Annex was published in February 2013 and entered into force on November 14, 2013. An Annex dedicated to safety management was developed to reinforce the role played by the State in managing safety at state level, stressing the concept of overall safety performance in all domains, in coordination with service providers.

4 SMSs of Civil Aviation Organizations Today

4.1 Results of the Survey «Safety First or Safety Sometimes?»

Results of a survey conducted among 48 representatives of Polish civil aviation organizations are presented in this chapter. The survey was sent to different types of Polish organizations: aerodromes, air navigation service providers, approved training organizations, aero clubs, maintenance and repair organizations, air operators and to continuing airworthiness management organizations. The survey consists of 14 multiple choice questions and an «open text» field. 30 safety managers or safety officers responded to the survey (4 of them have another job at the same time). The other respondents are top managers, internal auditors, air traffic controllers, airport duty officers, pilots, mechanics and other operational personnel.

The first two questions of the conducted survey were:

1. Does the top management of your organization know and understand just culture policy?
2. Does the top management of your organization ensure just culture?

For the two questions there were 35 'yes' answers and 13 'no' answers, meaning 27% of them were negative. Curiously, the answers for two questions from one respondent are not always the same. Table 2 illustrates the distribution of answers.

Table 2. Distribution of answers for the survey questions: '1. Does the top management of your organization know and understand just culture policy?' and '2. Does the top management of your organization ensure just culture?'.

Does the top management of your organization know and understand just culture policy?	Yes	No	Yes	No
Does the top management of your organization ensure just culture?	Yes	No	No	Yes
	33	**11**	**2**	**2**

It is hard to explain the cases of top management ensuring just culture but not understanding it – the Author considered these answers as the statistical error. However every 'no' answer to these questions is alarming and informs about an organization with no functioning just culture. Just culture is one of the most important principles of the SMS promoting the reporting of incidents, which enables to learn about causes of hazardous situations [13]. The presence of negative answers to the questions presented above means that people at work do not feel free to work, speak up and report harmful situations, conditions, events, incidents or accidents. It reduces cooperation, trust and reporting, prevents innovation, and adversely affects healthy judgements about risks that are part of everyday work [8]. Another question asked was «Did you experience a situation when a safety measure that you suggested was declined because of the lack of willingness of your colleagues?». The purpose of this question was to diagnose the approach of the entire personnel of an organization to safety-related measures (Fig. 1).

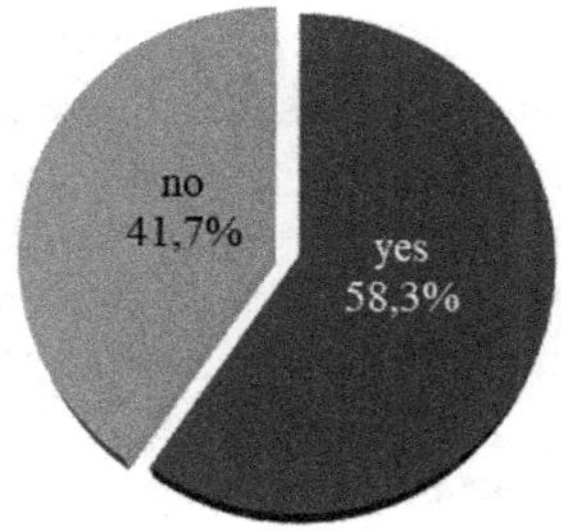

Fig. 1. Distribution of answers for the survey question: 'Did you experience a situation when a safety measure that you suggested was declined because of the lack of willingness of your colleagues?'

Almost 60% of respondents experienced a situation of the lack of contribution of their colleagues to safety management. It indicates a major difficulty for the SMS: without contribution (or at least without disturbance) of the entire community, it cannot act effectively.

Representatives of Polish civil aviation organizations were also asked about the status of the safety management in their firms. Almost 70% of answers defined the organizational unit responsible for the SMS as stand-alone department. Answering the next question of the survey, one quarter of the respondents admitted that safety is not the priority of the management in their organizations.

4.2 Top Management Contribution

The seclusion of the SMS is often related to the fact that this system is not profitable for an organization. It is obvious that the goal of top management is to ensure a high level of prosperity for the organization. If we consider a big organization, we have to take into account that top management is not necessarily involved personally in aviation. These people are there to manage a business. They support the SMS unit in the organization mostly because it is required by law. As long as they do not notice the potential benefits of the SMS, they will not treat it favorably.

In small organizations, the access to top management is easier, and managers are more often involved in aviation personally. That gives a chance that top managers know aviation history and the origins of the SMS. The most frequent problem of SMS implementation in such firms is the lack of funds and mismatch of the system design to the specifics of its activity. Both in small and big organizations, there is no money and no time for activities which do not bring benefits.

Taking a broad holistic view, safety management ensures development of aviation businesses. Thanks to the low rate of accident's passengers trust air transportation and are willing to use it. Different aviation organizations will experience decrease in connection with this. The costs of not managing safety of air operations can also be direct: compensations, reconstructions, increase of insurance premiums. The excellent reputation of a company may be also an argument in business negotiations. The widespread involvement of top management can be obtained by presenting potential benefits of a well-functioning safety management system.

Representatives of Polish civil aviation organizations answered another question of the survey: «My organization has enough staff to develop the safety management system». 28 of them answered «yes», the answers of another 19 respondents were negative. It is noteworthy that the safety management system can be integrated with other management systems of the organization. Integration saves time and money: adequately designed processes can be useful for managing such aspects as finance, personnel, quality, safety, security.

Considering the statistics presented below, we can assume that top management of most organizations sees some benefits of risk analysis and incidents investigation, and that there is will to improve the effectiveness of safety management (Table 3). However, the Author points out that there are some organizations in which the SMS is still at the stage of informal arrangements or at the stage of defining the processes.

Table 3. Distribution of answers for the survey questions: '8. In my organization, safety risks are analyzed to prevent incidents' and '9. Only events that are subject to mandatory reporting to the Civil Aviation Authority and the State Commission on Aircraft Accidents Investigation are investigated in my organization', '10. It is possible to monitor the effectiveness of the safety management system in my organization (we have effective tools for this)', '11. The safety management system in my organization is constantly being improved.', '12. Communication on safety matters with entities cooperating with my organization is open and effective'.

	Yes	No
In my organization, safety risks are analyzed to prevent incidents	**45**	**2**
Only events that are subject to mandatory reporting to the Civil Aviation Authority and the State Commission on Aircraft Accidents Investigation are investigated in my organization	**10**	**37**
It is possible to monitor the effectiveness of the safety management system in my organization (we have effective tools for this)	**36**	**11**
The safety management system in my organization is constantly being improved	**41**	**6**
Communication on safety matters with entities cooperating with my organization is open and effective	**35**	**11**

4.3 Personnel Involvement

Often employees are not engaged into safety management because they do not see its results. It lies deep in our human nature to demand a profit from any taken action: after clearly explaining what kind of involvement we expect from our personnel, we should present the potential benefits it would entail (benefits which may be noticed by them). The communication and transparency of the system are very important – a human acts better when they understand the goal of their activities.

When asked about the SMS, line employees often complain that the system is fancy, that it finds fault with everything and that safety personnel's eyes are prying. They are also anxious about comments of their colleagues. They doubt the just culture policy and they are overwhelmed by numerous procedures and documents.

In Eurocontrol's Just Culture Manifesto [8], it is explained that: «The organization must support people who are involved in or affected by accidents. This is the first priority after an unwanted event. Accidents can be traumatic experiences for all involved. People may be distressed or injured, physically or psychologically. Support for people is therefore the first priority after an unwanted event. While adverse events such as accidents are uncomfortable and often distressing experience, the learning process should not be. Safety investigations and organizational learning activities concerning unwanted events should – as far as possible – be positive experiences for all involved, improving the design of the system, helping individuals, teams and the organization to grow and become more resilient, and repairing – as far as possible –any damage done.»

To maximize personnel involvement, it is also essential to eliminate these barriers. To build a strong safety culture in the organization, consistency of actions and time are necessary – it is a long-term effort, which requires participation not only of organizations' staff members and managers, but also of state authorities. Maximum limitation of the documentation of the process will result in it becoming more practical; creation of simple, unambiguous procedures containing flow charts and the use of digital tools can make the process more accessible.

4.4 State Authorities

A strong, independent oversight constitutes the core of the safety management system. On account of that fact, civil aviation authorities are the subject of USOAP audits. Unfortunately, the Polish Civil Aviation Authority is experiencing shortages of resources. Many service providers in Poland complain about the insufficiency of the Authority. As industry's representatives are declaring, every CAA inspector can interpret the requirements differently, having a personal approach [12]. The president of the CAA reported in 2019 the dramatic situation of his bureau to the Polish government requesting additional funds. «This is the first case in Poland that one of the aviation organizations requested the German CAA to take over the oversight of its certification process. The situation is critical.», he said [5].

Also as an active regulatory body, which is open for the industry's needs, it plays a very important role in the SMS. As we read in a recent entry published on the blog of an aviation-related lawyer [6], the Ministry of Infrastructure, which is the Polish government agency responsible for aviation, is not always open and friendly towards the industry's representatives. Another problem indicated in the aviation society is the performance of the State Commission on Aircraft Accidents Investigation, which is a subsidiary body of the Ministry. The registries of occurrences have deficiencies and the classification of them is ambiguous. Questions about independency of the Commission, adequate staff and the lack of reports of some of the conducted investigations on Commission's website [4] are abundant.

All these factors lower the level of trust of Polish airspace users in aviation authorities. As we already witnessed in the history of aviation, weak state authorities have a huge, negative impact on the safety of air operations.

5 Conclusions

The Safety Management System of Polish aviation developed significantly, however there is still a long way ahead of us.

Significant differences of the development stage of the SMS among the organizations are present. It is essential that each organization tailors the implementation of safety management to fit their specific environment. Top management of aviation organizations not always notices the potential benefits of the SMS, which may cause the seclusion of this system. The Safety Management Team should be aware of operational issues and should be part of risk-based decision-making related to the organization's daily business. It is crucial not to treat the SMS as a stand-alone department or an undue burden. The results of the conducted research have confirmed that there are possibilities of solving emerging problems by improving the operation of Safety Management Systems in organizations.

Research has shown that the key element ensuring high efficiency of the process is to understand its essence and significance for maintaining an acceptable level of safety by representatives of the aviation industry: both by process contractors in civil aviation organizations and by the State aviation authorities.

Safety management processes will gain effectiveness thanks to:

- Education about history and development of the safety management system in aviation and other industries (involvement of the personnel, conscious management);
- Integration with other management systems (growth of database and of availability of resources; contribution of top managers);
- Limitation of documentation;
- Effective oversight and legislative process;
- Proper resources allocated to aviation State authorities.

The Author is aware of the complexity of the problem and plans to continue exploring it.

References

1. Annex 19 to the Convention on International Civil Aviation, 1st edn, ICAO, Montreal (2013)
2. Annual Activity Report 2018. European Union Aviation Safety Agency, Cologne (2019)
3. Annual Safety Review 2019. European Union Aviation Safety Agency, Cologne (2019)
4. Dańko, G.: PKBWL – instytucja we mgle? In: dlapilota.pl (2019). https://dlapilota.pl/wiadomosci/kontrola-ruchu-lotniczego/pkbwl-instytucja-we-mgle. Accessed 13 Apr 2020
5. Domański, A.: ULC walczy o fundusze, bo traci wydolność. In: rynek-lotniczy.pl (2019). https://www.rynek-lotniczy.pl/wiadomosci/ulc-walczy-o-fundusze-bo-traci-wydolnosc-6569.html. Accessed 13 Apr 2020
6. Doskocz, M.: Dość tej amatorszczyzny. In: latajlegalnie.com. https://latajlegalnie.com/2020/04/06/dosc-tej-amatorszczyzny/. Accessed 13 Apr 2020 (2020)
7. ICAO Homepage. https://www.icao.int. Accessed 08 Apr 2020
8. Just Culture Manifesto. SKYbrary.aero (2020). https://www.skybrary.aero/index.php/Just_Culture_Manifesto?fbclid=IwAR3HwmT9QKsG3mRDIvwptgtYThlExicgpgTwPFov622Z5mrae3oJ-kjmIqc. Accessed 08 Nov 2020

9. Kozłowski, M.: COVID-19 – inherentne operacyjne zagrożenie bezpieczeństwa. Biuletyn Bezpieczeństwa w Lotnictwie Cywilnym 3(13)/2020. Warszawa (2020)
10. Królikowski, W.: Komu i po co jest ten SMS – głos na forum SMS. In: dlapilota.pl (2014). https://dlapilota.pl/wiadomosci/polska/komu-i-po-co-jest-ten-sms-glos-na-forum-sms. Accessed 11 Apr 2020
11. Królikowski, W.: Nasza droga do lotniczej Europy, czyli przygody urzędnicze z ustawą, GILC i ULC. In: dlapilota.pl (2013). https://dlapilota.pl/wiadomosci/dlapilota/nasza-droga-do-lotniczej-europy-czyli-przygody-urzednicze-z-ustawa-gilc-i-ulc--1. Accessed 13 Apr 2020
12. Marzec, D.: The management of change as part of the safety management system of selected civil aviation organisations. War Stud. Univ. Sci. Q. **115**(2), 60–76 (2019)
13. Regulation (EU) No 376/2014 of the European Parliament and of the Council of 3 April 2014 on the reporting, analysis and follow-up of occurrences in civil aviation, amending Regulation (EU) No 996/2010 of the European Parliament and of the Council and repealing Directive 2003/42/EC of the European Parliament and of the Council and Commission Regulations (EC) No 1321/2007 and (EC) No 1330/2007

The Impact of Covid-19 on Airport Operations

Malwina Okulicz(✉) and Paulina Rutkowska

Faculty of Transport, Warsaw University of Technology, Warsaw, Poland
m.okulicz@wt.pw.edu.pl, paulina.rutkowska@pw.edu.pl

Abstract. The article deals with the impact of the Covid-19 pandemic on airport operations. The paper presents identified practicable solutions to reduce the risk of spreading the Covid-19 disease at the airport. The authors then determined how these solutions affect the implementation of airport operations, both in terms of the organization of the ground handling of the aircraft and the operational capabilities of the airport. As part of the work, observations and measurements of movements at the Chopin Airport were made during the suspension of passenger traffic and immediately after its restart. For this purpose, available data sources were used, i.e. the Airport Collaborative Decision-Making (A-CDM) platform for Chopin Airport, the FlightRadar24 website, as well as conducted field measurements. Collected data was used to perform a statistical analysis supplemented with a comparative research of selected operating parameters.

The first and key conclusion from the conducted research is the lack of appropriate procedures or operational readiness in the unprecedented situation of almost complete suspension of air traffic. The resumption of flights is associated with the reorganization of the airport's operational activities with the simultaneous implementation of new solutions during the pandemic. This results, among others, from the need to increase the automation of the information exchange in order to streamline and ensure the continuity of service processes in changing operating conditions. These observations designate the use of the potential of the existing A-CDM information exchange system, which generally has a positive effect on the course of airport operations through real-time data exchange. This system can be extended with an additional Covid-19 component and supply the obtained data to the entire air transport network.

Keywords: Airport operations · Airport data · A-CDM · Covid-19 · Traffic recovery

1 Introduction

Until now, aviation, apart from various market perturbations caused by events such as economic crises, natural disasters and political conflicts, has seen continuous growth. Since the beginning of aviation, no measure has been developed in the event of a global air traffic suspension. At the end of the first quarter of 2020, the Covid-19 pandemic outbreak became a fact, which almost completely suspended passenger air traffic around the world. This unprecedented event, unpredictable and unprepared for, froze the airline industry for almost 4 months. According to ACI estimates (as of 06/06/2020), airports

© The Author(s), under exclusive license to Springer Nature Switzerland AG 2021
A. Kwasiborska et al. (Eds.): ATE 2020, LNITI, pp. 86–105, 2021.
https://doi.org/10.1007/978-3-030-70924-2_8

lost more than 50% of passenger traffic and 57% (over USD 97 billion) of revenues from air operations in 2020 compared to foregoing operations [1].

Recent events related to the outbreak of the pandemic have shown how much impact it has on the smoothness and continuity of air traffic [15]. This situation has exposed many shortcomings in terms of preparing aviation for a situation of complete stoppage of air traffic. In addition, there were further unexpected issues that had not previously been considered, namely:

- Gradual recovery of air traffic (traffic recovery),
- Integrating the imposed restrictions and sanitary regime with the existing operational procedures and examining their impact on the implementation of the ground handling of the aircraft.

With time, links were reestablished, first domestic, then international - after gradual reopening of borders. The lack of procedures and unclear guidelines caused disarray at many airports. Limitations in terminals caused by physical distancing and the need to adapt the airport to the new rules showed the need to reorganize airport operations. These events showed how important proper organization and communication at the airport is. Airport Collaborative Decision-Making (A-CDM) is one of the many systems supporting airport operations. The implementation of airport systems such as A-CDM greatly facilitates and accelerates the exchange of information in the air transport network and supports the management of material and human resources at the airport.

The article focuses on examining the impact of the imposed restrictions and sanitary regime on the implementation of airport operations, using the A-CDM system for the exchange of information in the times of Covid-19, due to the undoubted potential of such systems to maintain traffic flow at airports.

In order to estimate the impact of the pandemic on airport operations and at the same time indicate potential of the existing A-CDM airport system to support ground handling processes, we developed the concept of the article as follows: first, we conducted a literature review on the impact of Covid-19 and the sanitary regime on aviation. Although the work focuses on airport traffic, the review is quite general due to the limited research and publications on airport operations.

In Sect. 3, we presented possible solutions for physical distancing that affect the operational work of the airports. Then, in Sect. 4, we analyzed the potential impact of the pandemic and the resulting limitations on aircraft ground handling. Section 5 presents the analysis of the airport's operational capabilities under Covid-19 conditions. Due to the need of providing appropriate operational conditions at the airport, within restrictions and new airport procedures, it is necessary to use solutions facilitating the cooperation of air traffic services, aircraft operators, air carriers and airport personnel. In the article, we indicated as an example the existing Airport Collaborative Decision-Making system, which is an integrator in terms of operational data exchange. In Sect. 5.1 we presented its characteristics, then in Sect. 5.2 we analyzed its functionality and in Sect. 5.3 we described the potential of the system under operational restrictions. As part of the research, we assessed the impact of changes in the airport operational activities, with an emphasis on the turnaround times of aircraft and the processes related to departure, in relation to the resumed air traffic.

2 Literature Overview

A modern airport is a complex environment in terms of size, traffic, safety requirements and operational procedures, the primary role of which is to create favorable conditions for the handling of passengers and goods by means of aircraft. All airport resources constitute an essential medium to conduct ground handling processes.

Ground handling is a basic part of airport operations. It includes all the services provided at the airport to the air carriers [2]. Ground handling consists of the following service categories [32]:

- Passenger service,
- Luggage service,
- Handling of goods or mail,
- Apron service,
- Aircraft service,
- Service for the supply of aircraft with propellants,
- Aircraft maintenance service,
- Support for operations and administrative activities related to the crew,
- Ground transport,
- Handling of in-flight supplies for aircraft.

The complexity of ground handling processes at an airport generates various challenges related to the requirements providing the safety of all components involved in it, but also the appropriate coordination of ground handling activities performed in parallel, often independently of each other.

So far, airports have faced many challenges, the main ones being [33]:

- Various types of acts of terrorism,
- Risks related to explosion or fire,
- Control of abandoned luggage,
- Access control to restricted areas,
- The presence of unmanned aerial vehicles,
- Cybersecurity.

The areas identified above cover the local environment of the airport. However, they do not contain the global problem, such as the ongoing Covid-19 pandemic. Undoubtedly, its effects can be clearly observed in aviation. The new threat in this context, concerning the spread of the virus by air, is an aspect that has been raised incidentally so far. The complete suspension of passenger air traffic in many parts of the world has not yet taken place so far, but it has allowed for the development of new solutions adapting airports to new service conditions.

After the outbreak of the Covid-19 pandemic, international aviation and sanitary organizations have undertaken the task of developing procedures and guidelines on a global scale, adapting the sanitary regime requirements to the implementation of ground operations. A number of guidance and bulletins have been created, which are updated on an ongoing basis with the development of knowledge and the results of progressive scientific research on the identified virus.

Guidelines presented by the International Air Transport Association (IATA) concerning ground handling [17] provide a quick reference to the various governmental and industry materials and updates on the Covid-19 outbreak specifically related to the ground handling industry. According to the document, the principles of most operational procedures are unchanged, while cleaning, good hygiene measures and consistent use of appropriate personal protective equipment (PPE) is recommended. The document presents:

- Recommendations for ground handling by key operational areas,
- Procedures that provide a quick reference and general guide for formally taking ground service equipment (GSE) Out of Operational Use (OOU), how to manage it while it is OOU, and the steps to return it to service,
- General guidelines for ground handling employees training management.

Another document developed by IATA is the Guidance on the Resumption of Operational Ground Handling [18]. It provides guidance on return to service after the partial or total closure of airline operations caused by the Covid-19 pandemic. The document includes considerations for risk assessment as well as highlighting various operational factors and safety risks which should be taken into consideration to ensure a safe and effective return to service.

Airport Council International (ACI) released a publication [1] which focuses on best practices and guidance tailored for both the initial phases of this return to business (restart) and the longer-term recovery processes of aviation operations. The document indicates possible procedures for securing airport staff and discusses issues related to both the direct operation of the airport during the functioning of the sanitary regime, as well as procedures aimed at risk mitigation: communication methods, determining changes, isolation of critical areas and areas not requiring external contact, setting up new services and designating new duties, etc.

In turn, the European Union Aviation Safety Agency (EASA) in cooperation with the European Center for Disease Prevention and Control (ECDC) issued a document [7] discussing operational guidelines for handling passengers and staff in the conditions of the Covid-19 pandemic. The material was established to help airports prepare for the return of regular air traffic.

The rapid spread of the virus requires frequent disinfection of places where larger groups of people are present. According to the IATA Medical Advisory Group [14], the aircraft cleaning and disinfection procedures, in excess of the previous norms, are likely to form part of the range of measures required in a restart process. It is likely that improved routine cleaning could provide reassurance to passengers and increase their confidence in the reduced risk of the transmission of communicable diseases in aviation processes. The purpose of IATA guidance is to provide airlines and ground handling providers with standardized procedures for cleaning and disinfecting of the cabin, flight deck and cargo compartments while taking into consideration the operational impact, cost and the practicality of mitigation measures for pandemic management.

Similar guidelines have been issued by EASA [6] giving proper consideration to the publications of the World Health Organization (WHO) and the ECDC in order to provide support to its stakeholders in terms of how cleaning and disinfection are expected to be performed.

The International Civil Aviation Organization ICAO has published a document [20] presenting measures to reduce the risk of spreading the virus divided into four modules, i.e.: airports, aircraft, crew, cargo. The airport module contains specific guidance addressing elements for: airport terminal building, cleaning, disinfecting, and hygiene, physical distancing, staff protection, access, check-in area, security screening, airside areas, gate installations, passenger transfer, disembarking, baggage claim and arrival areas. The aircraft module contains specific guidance addressing, among others, boarding processes, seat assignment processes, baggage, crew protection, management of sick passengers or crew members, and cleaning and disinfection of the flight deck, cabin, and cargo compartment.

The ICAO developed a dedicated webpage [22] related to the Covid-19 pandemic providing utilizable guidance and information. The Air Traffic Management (ATM) section supports collaboration decision making between all involved stakeholders for the efficient coordination and management of traffic during the Covid-19 disruption. The provided guidance is expected to support a smooth and less challenging return to normal operations. Additionally, potential hazards review, related to recovery to normal air traffic services (ATS) is presented to support ATS providers in their safety risk assessment processes.

However, it should be noted that there are currently few scientific publications dealing with the impact of Covid-19 on aviation. The effect of the introduced restrictions related to air travel on the passenger transport market and socio-economic development was the subject of considerations in the work of Iacus et al. [16]. On the other hand, publications [12, 17, 18] are related to physical distancing methods and measures limiting the spread of the virus at airports. Serrano and Kazda [32] attempted to determine the impact of the pandemic on the development of airports in post-pandemic times based on the identified development scenarios. The authors suggested solutions both for use at the airports in order to reduce the spread of the disease (mainly based on physical distancing) and to limit the negative impact of the pandemic on the financial situation of aviation industry companies.

The authors of the publication [24] further detailed the impact of Covid-19 on the functioning of airports analyzing the critical process of passenger handling service - security control. As part of the research, a simulation model of the operation of the security control lane in the COVID-19 epidemic was developed for various scenarios, comparing different ways of passenger flow. The aim of this article was to check the influence of physical distancing on the operation of security control stations.

The deficit of scientific papers on the impact of the Covid-19 pandemic on airport traffic can be justified primarily by the concentration of international aviation organizations and research centers on the impact of the virus on the safety and health of passengers and staff. Moreover, gradual resumption of air traffic occurred, depending on the country, at the turn of June and July 2020, so most of the investigations have only just started. One of the examples is this article, which presents the state-of-the-art research concerning the measurements and results of traffic observations at Chopin Airport during the suspension of passenger traffic and immediately after its resumption.

The current epidemiological situation is dynamic therefore it is necessary to constantly monitor the state of emergency and changes in regulations and procedures, which may significantly modify the air traffic profile in the coming months, or in the worst case, suspend it again.

Although the analyses presented in this article are the first attempts to address the issue, they were carried out in many aspects, in view of to the analysis of both the period of resumption of traffic and the ongoing restrictions, as well as the practically complete suspension of passenger operations.

3 Physical Distancing Solutions

According to the WHO [13] and Centers for Disease Control and Prevention (CDC) [25] physical distancing is a set of methods or measures aiming to prevent the transmission of a contagious disease by maintaining a physical distance between people and reducing the time in proximity with other people. The initially used term "Social Distancing" was changed by the WHO to "Physical Distancing" in order to emphasize the significance of maintaining space between people to avoid the spread of the virus.

This section shows a collection of the main solutions we gathered on the basis of the documents presented in the literature review. These recommendations, which focus on physical distancing, are crucial to develop a set of new instructions and rules ensuring airports operability at an identifiable level.

Physical distancing solutions have been introduced at most airports to ensure the safety of passengers and staff. The restrictions in the terminal that have a direct impact on ground operations are:

- Allocating every second stand - this applies to all terminal processes, i.e. check-in (both traditional and self-service), access control and security control, document control and gates,
- Separation of passengers in the terminal - including limiting the number of people in queues on passenger stairs or platforms and at gates,
- Controlled boarding - calling groups of passengers on board, according to the assigned seats in the aircraft.

The restrictions introduced in the process of ground handling of the aircraft, which extend the turnaround times of aircraft, are as follows [6, 10, 16–19, 23]:

- Suspending passengers on board - avoiding congestion in the baggage claim area by unloading and transporting checked baggage before the passengers arrive at the terminal, and by performing sanitary checks on passengers before disembarking,
- Reducing the number of passengers on buses - increased use of buses reduces resource availability and increases the demand for operating personnel,
- Controlled deboarding - calling groups of passengers to disembark in one communication route, according to the assigned seats in the aircraft,

- Disinfection of aircraft and shuttle buses - the use of complementary disinfection with additional methods, such as electrostatic spraying, UV treatment, ionization, fumigation, gaseous ozone etc. (although some of these methods are currently in the testing process, until positive and effective results are demonstrated and no long term effects on the aircraft structure and cabin are exhibited, they are not considered as an industry standard),
- Limitation of hand luggage - extension of loading/unloading luggage, due to the increase in the number of checked baggage.

4 Impact of Covid-19 on Ground Handling

The new regulation of physical distancing caused by the Covid-19 pandemic results in reorganizing the work of airport personnel. Due to the limitations caused by the change of airport procedures and the reduction of operations performed at the airport, there is a decreased demand for handling agent personnel. At the same time, when a virus infection is detected among employees, it is necessary to quarantine the entire shift, which significantly reduces the work of the ground handling agent and requires more flexibility and shift work.

One should consider the increased effort of handling agents in the area of passenger traffic organization, namely in the aspect of:

- Coordination and queuing to/from the aircraft,
- Ensuring everyone can maintain a physical distance,
- Transport of passengers.

From the point of view of the availability of airport vehicles, the demand for their number is increasing, due to the limits imposed on the number of passengers transported by shuttle buses between remote stands and the terminal. On the other hand, during the restricted air traffic phase, a reduced number of operations and a diminished load factor may balance this demand. Therefore, during periods of congestion, flexibility of the services should be ensured, e.g. by enabling the service of cruises by various agents, whose resources are used at different times during the day.

As a result of the introduction of physical distancing restrictions, first of all, ground handling processes directly related to passenger handling are changing. Passengers boarding and deboarding times will be extended (in particular at remote stands) as well as the occupancy of passenger buses.

The extension may also apply to the processing of checked baggage, due to the limitations of passengers' hand luggage resulting from general recommendations. Another important aspect when planning airport operations is to consider greater flexibility (buffer) in turnarounds, with respect to the necessity to perform procedures in case the infection is detected among passengers or cabin crew.

It should be taken into account that prolonged passenger processing times at the terminal and other additional processes, detailed in Sect. 3, may affect aircraft service times. On the other hand, extending the aircraft turnaround time may affect the airport's operational capabilities due to the lower availability of aircraft stands and airport equipment (GSE).

At the same time, the grounding of a large number of aircraft as a result of the suspension of most flights is associated with long-term occupancy of aircraft stands and additionally limits the capacity of the aprons. The release of parking spaces may take place by towing the machines, e.g. to pre-hangar stands, or by relocating them to less crowded airports.

Assessment of the airport operational activities, in particular the effects of changes in the implementation of ground handling processes, requires considering all the operational factors identified above.

5 Airport Operational Capacities

Airport performance is determined by its operational capabilities per unit of time. They must be adjusted to the preferences of air carriers, but also to local procedures and the resulting restrictions. This applies to both available airport resources (infrastructure, staff and equipment) as well as the declared capacity. The lack of experience of the managers in the situation of restoring the functionality of the airport after almost complete suspension of air traffic for a period of several months makes it possible to examine the system in completely new operating conditions, with relatively little, still limited traffic. Therefore, there are new issues to be clarified, including the crucial ones - how the restrictions related to the Covid-19 pandemic actually affect the course of turnarounds and how and with what measures the airport can rebuild its operational capabilities. The conclusions from the newly implemented physical distancing solutions indicate an even stronger emphasis on the cooperation of air traffic services, aircraft operators, air carriers and airport personnel. It is appropriate to first consider the already existing solutions rather than taking new, long-term initiatives, which as a result may not be necessary. Among the numerous existing systems and tools supporting operational processes at the airports, the choice of the Airport Collaborative Decision-Making (A-CDM), which acts as an integrator at the airport in terms of operational data exchange, is particularly appropriate. It may be helpful in assessing the airport's operational performance. The idea of the system is based on the completion of ground handling processes planned in advance at the appointed time. Therefore, it seems crucial to check the correctness and timeliness of completing the operational data in the A-CDM system concerning particular stages of ground operations. This will allow to determine the operational capabilities of the airport organism, both in terms of organization (adaptation of personnel to new operating conditions, effectiveness of A-CDM as a supporting system) and function (traffic profile, punctuality of operations).

5.1 Characteristics of the A-CDM System

The basic role of the A-CDM system is the real-time exchange of operational information between units responsible for aircraft handling. The factor enabling this process is the increasingly common automation of information flow. Despite the high level of automation, the concept is based on a human-centered system approach. It is aimed to support the implementation of human activities and, from an operational point of view, to schedule them over time. Flight is initiated by the activation of the flight plan in the information exchange system, which enables early planning of the allocation of resources, and further input of information about the assigned aircraft stand, taxi time from the runway to the assigned stand and vice versa, followed by the estimated time to exit the blocks and the planned time to start. Based on a variety of operational data from users of the information exchange system, it is possible to dynamically predict situations that may materialize during turnaround, such as an occupied aircraft stand or taxiway congestion. The key stage of the turnaround process is therefore the ground handling of the aircraft, the duration of which determines the accuracy, efficiency and predictability of decisions made regarding the course of subsequent stages of a flight. The full implementation of the A-CDM system follows the implementation of sixteen Milestones representing particular stages of a flight (see Table 1).

Table 1. A-CDM milestones (Source: own study based on [9]).

Milestone	Flight stage	Milestone	Flight stage
1	Flight plan activation	9	Deboarding
2	Flight data update	10	Ground handling service
3	Take-off from outstation	11	Boarding
4	Airborne	12	Ground handling ends
5	Final approach	13	Start-up request
6	Landing	14	Pushback
7	Taxi-in	15	Taxi-out
8	In-block	16	Take-off

For the purpose of information exchange, the airport uses the available technical measures at its disposal. However, it is required to develop software enabling the entry and management of data located in the AODB (Airport Operational Database), to which all authorized entities have access. Apart from the involved air traffic services, airport coordination, handling agents, airlines and aircraft operators, another important user of the system is the Network Manager - acting as an intermediary in European air traffic. The data available in the AODB are archived and distributed to users in real time. It aims to maintain situational awareness of airport services, but also of the entire air transport network in terms of planned traffic in the airspace and at destination airports. In a situation where this traffic changes dynamically, the global data distribution system becomes even more essential due to the need of adapting the operational capabilities to the expected traffic. Despite the fact that the concept of A-CDM was created to support the constantly expanding flight schedule, its functionality is also significant in the situation of the categorical limitation of the passenger traffic. **The extent to which the functionality and usability of the system was maintained during a pandemic is an important element of research and is a subject to evaluation in the following subsections of this article.**

5.2 Analysis of the Functionality of the A-CDM System

Basically, the A-CDM system fulfills its role in the case of congested airport traffic, with the overwhelming information load about flights and limited human perception. It may be presumed that when the flights are suspended, the usability of the A-CDM diminishes or its purpose alters. And when the flights are resumed, potential disuse of the system may somehow be renewed. Therefore, the functionality of the A-CDM system during the Covid-19 pandemic was analyzed in search of solutions on how the airport manages the resumption of traffic, facing the reduced operational parameters. The research was carried out on the example of the largest airport in Poland - Warsaw Chopin Airport. For this purpose, available data sources were used, i.e.:

- A-CDM platform for Chopin Airport [8] intended for aircraft operators and mobile users,
- FlightRadar24 website [11] - flight radar, and field measurements were carried out in the vicinity of Warsaw Chopin Airport.

The scope of the research included the A-CDM system observations and air traffic measurements:

a) Before the outbreak of the pandemic, with high traffic rates and when system implementation at the airport was at an advanced stage, but still in the testing and streamlining phase.
b) At the time the pandemic was announced, where passenger traffic was fully suspended, and only repatriation and cargo flights were executed.
c) During the pandemic, where passenger traffic has partially resumed.

Within bullet c. more detailed traffic studies were undertaken. The following operational parameters were analyzed:

- ALDT - Actual Landing Time,
- ATOT - Actual Take-Off Time,
- SOBT - Scheduled Off-Block Time,
- TOBT - Target Off-Block exit time,
- TSAT – Target Start-Up Approval Time
- CTOT - Calculated Take-Off Time.

The measurement test concerned 200 turnarounds performed from July 1st to August 12th, excluding Saturdays and Sundays, just after the resumption of international flight operations at Chopin Airport.

Pre-pandemic Air Traffic Analysis. Until recently, air traffic in Poland and in Europe recorded a continuous increase in passenger traffic. In 2019 the Chopin Airport handled over 180,000 flights, which indicated an approx. 5% increase in traffic compared to the previous year and as much as a 15% increase in relation to 2017 [27]. When analyzing the operational capabilities, the airport operated at the limit of its capacity and everything indicated the need to expand the airport infrastructure in response to the constantly growing transport demand. However, after the Covid-19 pandemic outbreak, the situation has changed dramatically, and the traffic in the first half of 2020 decreased by almost 60% compared to the corresponding period of the previous year. Detailed data is presented in Fig. 1.

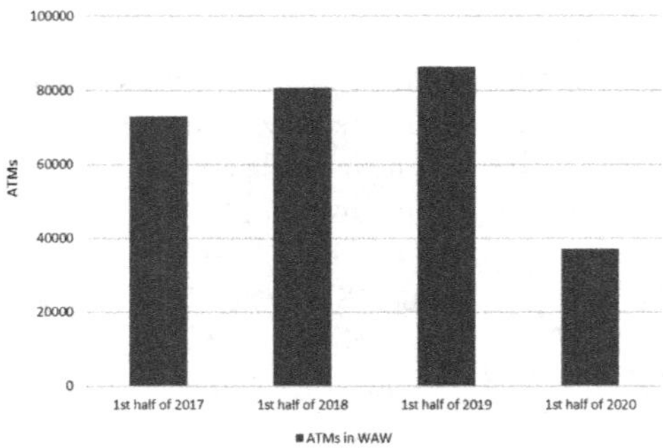

Fig. 1. The number of flights at Chopin Airport in 2017–2020 (Source: own study based on [3–5, 27, 28]).

In order to improve the operational activity of the airport in Warsaw, the Airport Collaborative Decision-Making System (A-CDM) was implemented in the first half of 2019. Due to the complexity of the system structure and the need to implement and master new operating procedures by its users (airport staff, air traffic services, handling agents, airlines and pilots), the implementation process had to be supported by a series of system functionality tests. With heavy traffic, areas which are subject to streamlining have been identified, so the A-CDM system was still being tested until the outbreak of the pandemic.

Traffic Analysis During Suspension of Passenger Operations. In aviation, the beginning of the Covid-19 pandemic was manifested by the introduction on March 15, 2020 [29] of a ban on domestic and international passenger flights to and from Poland, which had a clear impact on traffic statistics. The European Airports Association – ACI Europe recorded a 93% decrease in air traffic in June 2020, compared to June 2019 at European airports. The Warsaw Chopin Airport, between March and June, mainly carried out cargo and repatriation flights from various countries around the world. According to the airport's data, in June 2020, the airport carried out 1,200 flights, which means a decrease of over 92% compared to last year.

Due to limitations in airport traffic, it was easy to observe the airport's operational activity and identify typical trends. First, we noticed an irregularity in completing A-CDM procedures. The main observations are the frequent lack of data updates on the A-CDM platform or simply not respecting its content. As soon as it was ready for departure, the aircraft left the parking stand, which took place even an hour before the estimated TOBT time. Probably with a few or a dozen operations a day, it does not disturb the airport's operational capabilities, nor does it generate organizational problems. However, the above conclusions show the untapped potential of the A-CDM system.

Analysis of Resumed Air Traffic During the Pandemic. The Chopin Airport began the gradual resumption of domestic flights already in June 2020 [30], and international operations from July 1, 2020 [31]. The summary of air traffic at the Warsaw airport since the pandemic was announced is presented in the Fig. 2. The chart also shows the percentage drops in air traffic compared to last 2019.

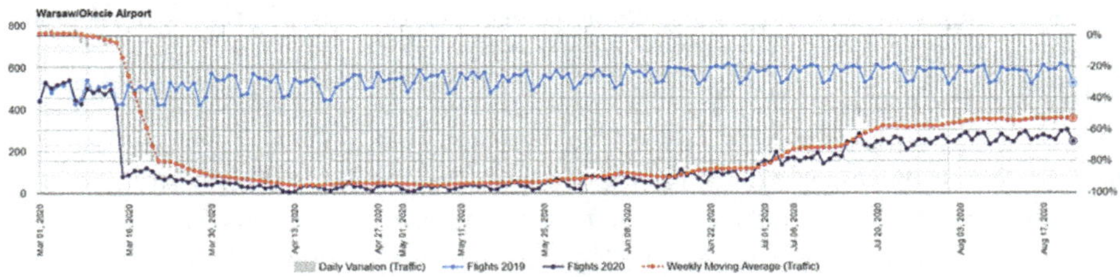

Fig. 2. Daily traffic statistics at the Warsaw Chopin Airport. (Source: [26])

Due to the relatively low traffic in the investigated airport, it was possible to conduct further observations of the airport traffic, paying particular attention to the organizational aspect based on the operation of the A-CDM system supporting ground handling processes at the airport.

Initially, on the basis of the conducted research, the operational factor of traffic was analyzed. The measurements were primarily aimed at examining the aircraft turnaround times. About 75% of measurements were taken in the afternoon and evening hours, while the morning measurements happened to be mostly overnights and it would be difficult to determine the turnaround time trends based on them. Analysis of the results of the undoubtedly indicates a significant change in the profile of the traffic (see Fig. 3), where short-haul flights predominate, while long-haul flights are organized only in the charter cruises formula.

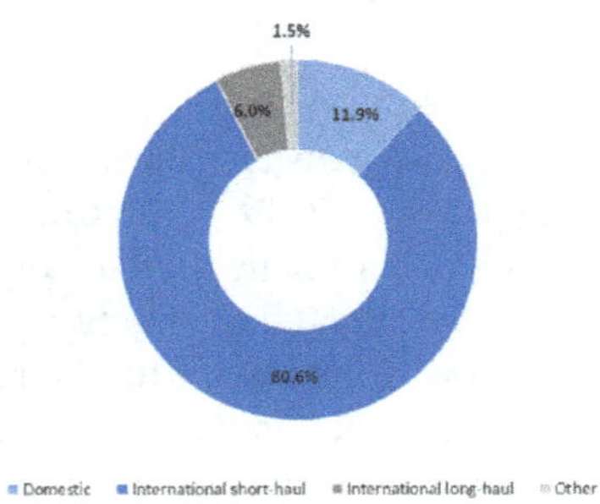

Fig. 3. Traffic profile at Chopin Airport in the analyzed period (Source: own study).

The lower domestic traffic was influenced by the greater confidence of passengers in making short trips by publicly available road or rail transport. On the other hand, short-haul flights dominated among international operations, which results mainly from reduced demand for travel and bans imposed on long-haul flights (USA, Asia, etc.). However, this ban did not include, inter alia:

- Operations from European Free Trade Association (EFTA) member states - parties to the agreement on the European Economic Area, the Swiss Confederation and European Union Member States, which translated into the dominance of short-haul flights,
- Flights, the so-called repatriation, which constituted a large part of the long-haul flights,
- Others, including those implemented for the needs of the Armed Forces of the Republic of Poland.

Other operations identified were short cruises over the airport-proximate area, possibly to test machines after long grounding or to maintain their airworthiness.

Subsequently, based on the observed traffic parameters, it was possible to determine the aircraft turnaround times, for which the average, minimum and maximum values from the set of results are presented in Table 2. The results from the same period in 2019 [26] were analyzed in order to compare the operational parameters from the previous year. For the comparison to be reliable, only the aircraft that were observed during the measurements carried out in 2020 were covered. The results presented in Table 2 include aircraft turnarounds carried out on the same day, excluding overnights.

Table 2. Analysis of aircraft turnaround times (Source: own study).

Value	Turnaround in 2019	Turnaround in 2020
Average	01:48:52	03:22:56
Minimum	00:30:00	00:42:00
Maximum	13:10:00	16:14:00
Average for wide body aircrafts	04:14:35	10:33:12
Average for narrow-body aircrafts	01:42:03	03:00:17

It can be seen that for all statistical parameters, the turnaround time has been extended compared to the corresponding period of 2019. The mean value of the turnaround time nearly doubled, which indicates a much smaller number of aircraft turnarounds during the day and thus a greater load on the airport infrastructure (longer occupancy of aircraft stands). Despite the significantly lower load factor for flights, the minimum turnaround time has also changed, the value of which increased by 40% compared to 2019. The extension of the turnaround time mainly results from the reduced number of operations performed, but also from the changes introduced in operational procedures during the pandemic. This is mainly due to the imposition of restrictions on performing the long-haul flights, which resulted in a reduction in the number of operations of this type of aircraft and their longer parking time. Table 3 shows what proportion of all completed operations were flights under 1 h and what was their average turnaround time.

Table 3. Analysis of aircraft turnaround times under 1 h (Source: own study).

Year	Number of operations < 1 h [%]	Mean turnaround time of operations < 1 h
2019	19%	00:47:37
2020	15%	00:51:04

Only 15% of operations from the 2020 research sample are flights with a turnaround time of less than 1 h, while in the same period of 2019 more, 19% of operations recorded a turnaround time shorter than 1 h. It shows that the mean turnaround time has increased while the number of operations has decreased significantly. It is especially noticeable for wide-body aircraft, which results from the small number of long-haul flights and their more frequent basing at the airport.

Subsequently, flights were analyzed for delays in departures and departures ahead of schedule. The mean values of both variants were assessed on the basis of slot adherence. The slot tolerance window determines the acceptable time frame within which the aircraft should take off, namely minus 5 min before CTOT, to plus 10 min after CTOT. Taking into account the fact that the CTOT time during the measurements was unavailable in the A-CDM platform, we made a comparative analysis of the actual ATOT time to the SOBT scheduled time. It required the calculation of the taxi-out time to the runway, which for the test sample was averaged to the value of 5 min.

Table 4 and Table 5 show the values of the mean delay and acceleration of departure, the number of aircraft in the slot and the average times only for flights within the slot window. More than 60% of the take-off operations were delayed from among the tested sample. The average time of this delay is presented in the first column in Table 4.

Table 4. Analysis of delays in relation to the schedule (Source: own study).

Average delay time	Amount of aircraft within the slot tolerance window [%]	Average delay time for flights within the slot
00:12:41	39	00:04:36

Fewer, i.e. slightly more than 20% of the launch operations from the studied sample took place prior the planned time. The average time ahead is shown in the first column in Table 5.

Table 5. Analysis of accelerations in relation to the schedule (Source: own study).

Average time ahead	Amount of aircraft within the slot tolerance window [%]	Average time ahead for flights within the slot
00:05:05	16	00:01:59

It can be seen that the average delay time exceeds the permitted time for slot tolerance window (from −5 to +10 min). Indeed, only 39% of the delayed operations and 16% of the operations were completed ahead of time within the slot tolerance window. This means that even with a relatively low traffic volume, but usually with an extended time, a large proportion of the operations were not performed within their allotted time. Perhaps the main cause of the identified deviations is the organization of airport work.

The organizational factor was assessed on the basis of observations of changes in the A-CDM platform [8] supported by observations of real time aircraft movements. In accordance with the parameters listed in Sect. 5.2, the platform provides access to four of them, namely:

- SOBT - scheduled time in accordance with the flight plan, available on the platform two hours ahead,
- TOBT - target off-block time, which appears automatically together with the confirmed SOBT time and is equal to SOBT,
- TSAT - appears at TOBT-40min and takes the value assigned to the pre-departure sequence algorithm,
- CTOT - appears after the TOBT is validated and is calculated automatically based on adding the estimated taxi-out time to the runway threshold to the TOBT.

Throughout the observation period of the platform, the SOBT time changed a single time for one flight, while the CTOT time appeared only once in the platform for one operation - in other cases it was unavailable or not entered. The TOBT time, crucial from the point of view of completing the ground handling processes, according to the guidelines of [9] could be updated any number of times after the TSAT time was determined - usually airports recommend changes up to 3 times. At the Chopin Airport however, it happened that it was updated even 4 times after this time (see Table 6).

Table 6. TOBT changes during observations (Source: own study).

No changes	1 change	2 changes	3 changes	4 changes
54%	37%	8%	1%	1%

The TSAT changes occurred accordingly to the TOBT alternations. Occasionally, during the departure peak, it was necessary to sequence the aircraft, and thus transmit the TSAT later. The consistency of both time values in the A-CDM system is assessed in Table 7.

Table 7. TOBT and TSAT compliance (Source: own study).

Compliant	Non-compliant
96%	4%

Verifying compliance with TOBT was difficult due to the lack of a full field of view (of each aircraft stand), or in the case of verifying operations on the FlightRadar website [11], relatively frequent loss of signal from the transponder and disappearance of the plane from the radar. Therefore, less than half of the measurements were tested, which allowed to state that at least 70% of the aircraft were leaving the stand in accordance with the [9] assumption, i.e. TOBT $\pm$ 5 min.

In the sample, for 25% of operations, earlier readiness for departure was declared, i.e. earlier TOBT time. The extreme values for the updated TOBT time concerned individual situations, while for the remaining majority the deviation oscillated around the mean value presented in the Table 8.

Table 8. Declaring earlier readiness for departure than TOBT (Source: own study).

Minimum deviation	Average deviation	Maximum deviation
00:01:00	00:09:47	01:05:00

Less, i.e. 20% of operations, declared later readiness for departure, which means that the TOBT time was delayed (see Table 9). The very long delay in terminating the ground handling operations affected individual operations but had a major impact on inflating the mean TOBT deviation. Many of the operations had a delay greater than or equal to 5 min.

Table 9. Declaring later readiness for departure than TOBT (Source: own study).

Minimum deviation	Average deviation	Maximum deviation
00:05:00	00:19:40	01:59:00

The information collected above is supplemented by additional observations collected during the measurements:

- Single turnarounds, despite the ongoing ground handling, disappeared from the platform and reappeared after a while - which may confuse the user.
- In isolated cases, the allotted TSAT was missing after the TOBT was updated, suggesting that operations were not included in the departure sequence.
- Single operations had a different destination in the A-CDM platform than on the radar [11].
- There have been occasions that operations have disappeared from the platform without a rational operational explanation - perhaps this suggests a flight cancellation or flight plan change.
- On the last day of measurements, the key - from the point of view of the purpose of the A-CDM platform - TSAT values for all operations were missing.

5.3 A-CDM Potential

The A-CDM system fulfills its role as long as the A-CDM platform is supplied with reliable operational data, in a timely manner and with sufficiently precise accuracy.

In order for these requirements to be met and for the A-CDM potential to be exploited, it is necessary to perform the operational procedures in accordance with a plan. One such procedure is the A-CDM procedure. The disarray caused by the Covid-19 pandemic is a proof that the system is susceptible to disruptions which leads to abandoning the operating measures taken to date in favor of adapting to completely new operating style.

Based on the research, it can be seen that with the outbreak of the pandemic and the restrictions of traffic at Warsaw Chopin Airport, the systematic and correct supplementation of data into the system was abandoned, which directly translates into the usability and further functionality of A-CDM. Perhaps, given the large restriction of traffic at the airport, the failure of the A-CDM system to operate correctly is not a major problem, although relatively low air traffic is an ideal circumstance for testing the operation and demonstrating the potential of the A-CDM.

As flight links are re-established and the number of operations increases, it is necessary to ensure an adequate level of communication to radicalize cooperation between air traffic entities. According to [21], an important need is to have a tool for the information exchange between adjacent Area Control Centers (ACC). The use of the simplified A-CDM procedure between air traffic control services of neighboring regions is extremely justified in the ongoing coordination of traffic flow, in dynamically changing airspace availability, usually at the country level. Increasing the automation of information exchange, e.g. between the pilot and the marshaller or the ground handling coordinator, will also improve the organization of work at the airport itself and reduce the problem of frequent direct contacts. Therefore, first of all, it is necessary to update the A-CDM platform with key data on the epidemiological situation at a given airport (or the airspace over the airport and adjacent areas), and secondly, to maintain an appropriate level of quality of the information provided, so that the users do not lose confidence in the correctness of the reported data, taking into account the experiences dictated by the outbreak of the pandemic.

6 Possible Solutions and Development Directions

The conducted literature review shows the insufficiency of relevant scientific research and analyses of the impact of the Covid-19 pandemic on the air traffic. The organization of the airport's work deserves special attention, as situations of traffic restrictions and constant changes in airlines flight links around the world cause uncertainty and disarray that negatively affects the operational continuity of aviation systems. Therefore, it is necessary to introduce procedures at airports and the use of automation tools that will streamline the operation of services and regulate surface movements. An example would be the A-CDM system that integrates all operational data. During the pandemic, the system could be additionally expanded with a Covid-19 module that would collect and distribute information about the epidemiological situation in the airport area including a list of new operating procedures. For example, in the event of an outbreak of Sars-Cov-2 virus in an aircraft, the A-CDM system, after an earlier alert, will distribute this information to all interested parties, who will be able to organize emergency services in advance, redirect the aircraft to a designated stand, plan extension turnaround time due to disinfection, or reorganize the work of the staff. The above module could easily transmit messages containing similar information to neighboring ACC stations. The following indicators may be helpful in defining the factors to be monitored:

a) Static

- List of countries banned from international flights,
- Airspace zones excluded from use for transit traffic,
- List of new or updated aerodrome procedures.

b) Dynamic

- Alert about the threat of infection with Sars-Cov-2 virus on a given flight,
- Aircraft occupancy - including a list of crew members, passengers and quarantined personnel,
- Airport zones and resources allocated to support a cruise at risk of contamination,
- Information on temporary separation or closure of airport zones from use due to the maintenance of physical distance.

7 Summary and Conclusions

The events of the first half of 2020 proved that it is possible to almost completely stop air traffic on a global scale. Restoring domestic and international flights requires even closer cooperation between countries based on the synchronization of actions taken. The new outbreaks of the Covid-19 pandemic cause individual and uncoordinated restrictions at the country level, which translate directly into airport operations, causing disarray in the airline industry and, moreover, decrease of passenger confidence in air travel. While the lack of coordination between countries in air travel management can be solved, for example, by introducing mandatory virus testing at all airports or by developing common epidemiological assessment criteria, improving cooperation and situational awareness at airports can be solved by using assistive devices (such as A-CDM) for the existing systems. The article analyzes the operational data collected during the pandemic at the Chopin Airport in order to investigate the impact of Covid-19 on the operational (airport operations) and organizational (human and material resources) factor.

The conducted research indicates that the pandemic affected the ground handling of aircraft to some extent, changing the profile of the traffic, which was adapted to the applicable restrictions and restrictions, and not to the real needs or preferences of passengers. In addition, aircraft turnaround times have almost doubled the average values of those from the same period in 2019. This is mainly due to the reduction in the number of flights performed. Another identified reason is the change of the operating procedures used so far during the implementation of the ground handling of the aircraft, including frequent disinfection and cleaning of the cabins, or longer boarding, deboarding of passengers due to the need to maintain social distance. When traffic is limited it may not be a big issue, however as the number of operations increases, the need to optimize maintenance activities will also increase. In order to maintain the airport's operational capabilities at an appropriate level, as well as to mitigate the disarray caused by the Covid-19 pandemic in the entire air transport network, it seems reasonable to use the A-CDM system for these purposes.

The main advantage of A-CDM is the ability to dynamically manage aircraft operations and the resources necessary to handle them, due to the database updated by users on an ongoing basis. It should be emphasized, however, that in order for the system to completely fulfill its role, it is necessary to ensure its continuity. The research carried out at the Chopin Airport showed that the functionality of the system was partially abandoned during the suspension of air traffic, and later during its gradual recovery. The time of the pandemic seems to be the right moment to use the potential of the system and implement new solutions. From the point of view of the unstable epidemiological situation that has a strong impact on aviation, we recommend implementing in the A-CDM system, inter alia, a module that allows the air transport network to be supplied with information about new restrictions at the airport, but also information about the temporary exclusion of airspace zones or airport zones, or detailed information about possible Sars-Cov-2 infection on a given flight.

References

1. ACI: Aviation Operations During COVID-19 Business Restart and Recovery, Issue 01 (2020)
2. Art. 176 of the Aviation Act of 3 July 2002 (Journal of Laws of 2019, item 1580)
3. Chopin Airport, News and events. https://www.lotnisko-chopina.pl/pl/aktualnosci-i-wydarzenia/0/1004/szczegoly.html. Accessed 10 Aug 2020
4. Chopin Airport, News and events. https://www.lotnisko-chopina.pl/pl/aktualnosci-i-wydarzenia/0/993/szczegoly.html. Accessed 10 Aug 2020
5. Chopin Airport, News and events. https://www.lotnisko-chopina.pl/pl/aktualnosci-i-wydarzenia/0/997/szczegoly.html. Accessed 10 Aug 2020
6. EASA: Guidance on aircraft cleaning and disinfection in relation to the COVID-19 pandemic. Issue 02 (2020)
7. EASA, ECDC: COVID-19 Aviation Health Safety Protocol Guidance for the management of airline passengers in relation to the COVID-19 pandemic. Issue 01 (2020)
8. EPWA, A-CDM Platform. https://pilot.polish-airports.com/. Accessed 14 Aug 2020
9. EUROCONTROL: Airport CDM Implementation - The Manual. Version 5.0 (2020)
10. EUROCONTROL: Impact assessment of covid-19 measures on airport performance, Ed. 1.0 (2020)
11. FlightRadar24. https://www.flightradar24.com/. Accessed 14 Aug 2020
12. Habida, E., Ebrahim, S.H., Gozzer, E.: COVID-19: facemask use prevalence in international airports in Asia, Europe and the Americas, March 2020. Travel Med. Inf. Dis. **35**, 101637 (2020)
13. Harris, M., Adhanom Ghebreyesus, T., Liu, T., Ryan, M. J., Vadia, Van Kerkhove, M. D., Diego, Foulkes, I., Ondelam, C., Gretler, C.: Costas: COVID-19. World Health Organization (2020)
14. Nakamura, H., Managi, S.: Airport risk of importation and exportation of the COVID-19 pandemic. Transp. Policy **96**, 40–47 (2020)

15. Iacus, S.M., Natale, F., Santamaria, C., Spyratos, S., Vespe, M.: Estimating and projecting air passenger traffic during the COVID-19 coronavirus outbreak and its socio-economic impact. Safety Sci. **129**, 104791 (2020)
16. IATA: Aircraft cleaning and disinfection during and post pandemic, Ed. 1 (2020)
17. IATA: Guidance for ground handling during COVID-19, IATA, Ed. 5 (2020)
18. IATA: Guidance for ground handling return to service, IATA, Ed. 1 (2020)
19. ICAO: CART Report: Flightplan for Restart and Recovery (2020)
20. ICAO, Council Aviation Recovery Task Force: Take-off: Guidance for Air Travel Through the COVID-19 Public Health Crisis (2020)
21. ICAO: Covid-19 outbreak simplified procedure for air traffic management collaborative decision making and sharing of information, Ed. 1.0 (2020)
22. ICAO, COVID-19 Safety Operational Measures. https://www.icao.int/safety/COVID-19OPS/. Accessed 10 Aug 2020
23. ICAO: Effects of Novel Coronavirus (COVID-19) on Civil Aviation: Economic Impact Analysis. Montréal, Canada (2020)
24. Kierzkowski, A., Kisiel, T.: Simulation model of security control lane operation in the state of the COVID-19 epidemic. J. Air Transp. Manage. **88**, 101868 (2020)
25. Kinlaw, K., Levine, R.J.: Ethical guidelines in Pandemic Influenza - Recommendations of the Ethics Subcommittee of the Advisory Committee to the Director, Centers for Disease Control and Prevention. Centers for Disease Control and Prevention (2020)
26. OAG Schedule Analyser. Accessed 01 Sep 2020
27. Polish Civil Aviation Authority, Number of passengers and operations performed in domestic and international traffic - regular and charter in 2017–2019. https://www.ulc.gov.pl/pl/. Accessed 26 Aug 2020
28. Polish Civil Aviation Authority, Number of passengers and operations performed in domestic and international traffic - regular and charter in the first quarter of 2018–2020. https://www.ulc.gov.pl/pl/. Accessed 26 Aug 2020
29. Regulation of the Council of Ministers of March 13, 2020 on air traffic bans (Journal of Laws of 2020, item 436) (2020)
30. Regulation of the Council of Ministers of June 16, 2020 on air traffic bans (Journal of Laws 2020, item 1050) (2020)
31. Regulation of the Council of Ministers of 30 June 2020 on air traffic bans (Journal of Laws 2020, item 1160) (2020)
32. Regulation of the Minister of Transport, Construction and Maritime Economy of 19 November 2013 on ground handling at airports (Journal of Laws 2018.1287) (2013)
33. Serrano, F., Kazda, A.: The future of airports post COVID-19. J. Air Transp. Manage. **89**, 101900 (2020)

Modelling of Runway Capacity for the International Airport Krakow-Balice Using Aimsun Next Software

Anton Pashkevich(✉), Jakub Dyrcz, and Olaf Dubiel

Politechnika Krakowska, ul. Warszawska 24, 31-155 Kraków, Poland
apashkevich@pk.edu.pl, jakub.dyrcz.pk@gmail.com,
olafdubiel20@gmail.com

Abstract. During the last years, different producers of simulation software – PTV AG, INCONTROL Simulation Software, Aimsun company – turned more of their attention to the problematics of air transport and, especially, to simulations at the airports. It was connected with the development and improvement of new tools to model pedestrian flows, which first of all allow to simulate airport passenger flows. Unfortunately, at the moment the process remains in this point. Existing tools to create models of, for example, runway capacity, airport apron capacity or air routes still require improvement. Within the scope of this research work, an attempt was made to model the runway of the Krakow-Balice International Airport using Aimsun Next software with the aim to assess its capacity. The results showed the general feasibility of using such a simulation tool. The advantages, disadvantages as well as limitations of selected software were also described.

Keywords: Runway capacity · Air operation · Arrival · Departure · Simulation modelling · Aimsun Next

1 Introduction

During the last decade, air transport became one of the most popular, fastest and safest forms of travel around the world. Covering the needs and expectations of people and organizations, it increased the load put on the air transport infrastructure, which resulted in a number of problems such as, for example, lack of slots in large airports. In such a situation, the right assessment of airport infrastructure opportunities could be a good approach, which could support the organizational or investment decision-making process. The simulation tools can make this process easier. In the case of the runway capacity estimation, a number of different special software packages is mainly utilized: Total Airspace and Airport Modeller (TAAM) [1], Airport Capacity Analysis Through Simulation (ACATS) [2], Airfield Capacity Model (ACM) and Runway Simulator (rS) [3], Visual SIMMOD [4], *runway*SIMULATOR [5], Arena [6], etc.

The purpose of this article is to assess the maximum hourly runway capacity at the Krakow-Balice International Airport and further to verify it by using such simulation tools as Aimsun Next software.

© The Author(s), under exclusive license to Springer Nature Switzerland AG 2021
A. Kwasiborska et al. (Eds.): ATE 2020, LNITI, pp. 106–115, 2021.
https://doi.org/10.1007/978-3-030-70924-2_9

2 Runway Capacity Calculation

2.1 Attempts to Calculate the Runway Capacity

There is a number of literature sources describing attempts to calculate and to optimize the runway capacity. The research works were based on various conditions and different airports.

Pavlin and co-authors calculated the runway capacity using actual time of take-off and landing, then the idea to improve the capacity was based on the reduction of the runway occupancy time (ROT) [7]. Another approach considered the separation between aircraft giving them the primary role [8–10]. It must be mentioned that the issue of separation was pointed out as a self-contained problem to solve. Besides the typical calculation, there were also attempts to develop an analytical model to analyze the runway capacity [11]. A more detailed review concerning runway capacity assessment as well as the factors influencing its performance was presented by Safrilah and Putra [12].

As it was mentioned above, simulations of the runway were done mainly based on special software packages. According to our knowledge, the products of such popular companies as PTV AG, INCONTROL Simulation Software, Aimsun Next software had not been previously applied to solve such issue as the runway simulation.

Within the scope of this article, the approach described by Grebenšek and Pavlin in 2004 was chosen to calculate the runway capacity for the Krakow-Balice International Airport [13]. The calculated outputs such as the number of arrival and departure operations will be a base to create a flight schedule for the simulation model.

2.2 Capacity of the Runway at the Krakow-Balice International Airport

Table 1 presents the input data required to provide the calculation of the runway capacity using selected method. It includes information for each category of aircraft concerning its landing speeds and ROT as well as data on the percentage share of each aircraft category in all air operations performed during the analyzed day. ICAO indicates a specific speed range in knots for each aircraft category to be achieved when performing a certain maneuver or during a specific approach phase [14].

Table 1. Input data to calculate the runway capacity.

Aircraft categories	Landing speed [kn]	ROT [s]	Percentage of current aircraft category	
			Arrivals	Departures
B	140	100	1%	1%
C	140	80	0%	0%
D	140	70	75%	75%
E	130	60	22%	22%
F	120	50	2%	2%

Source: own work based on ICAO, 2006

The percentage shares were determined based on the data on the number of landing aircrafts and their types from the whole day on 07.06.2019. The Krakow-Balice Airport accepted only one airplane of category B per day and three times a week in the summer connection network. This is the reason of their small share. Aircrafts from category C, such as Boeing 757, rarely land at the analyzed airport: they do not occur on regular routes. The biggest group is represented by aircrafts from category D (e.g., Boeing 737, Airbus A320). They are used by airlines such as Ryanair, EasyJet and Lufthansa, which perform the largest part of air operations to and from the Airport Krakow-Balice.

Calculation results are as follows: the runway capacities for arrivals only and for departures only are 23 operations/hour and 31 operations/hour, respectively.

To obtain the maximum capacity for departures in the mode of mixed operations, all the values in Table 2 must be summed, which returns the value of 22 operations/hour. This value is much smaller than the runway capacities for departures only. The amount of arrival operations stayed the same as it was calculated – 23 operations/hour. As a result, the maximum capacity for mixed operations can be 45 aircraft operations per hour.

Table 2. Amount of departures for a certain combination.

Aircraft categories	B	C	D	E	F
B	0,00	0,00	0,00	0,05	0,00
C	0,00	0,00	0,00	0,00	0,00
D	0,17	0,00	12,72	3,73	0,34
E	0,05	0,00	3,73	1,09	0,10
F	0,00	0,00	0,34	0,10	0,00

Source: own work

Based on the calculations made, Fig. 1 has been created, which presents the maximum capacity for air operations: arrivals, departures and mixed ones.

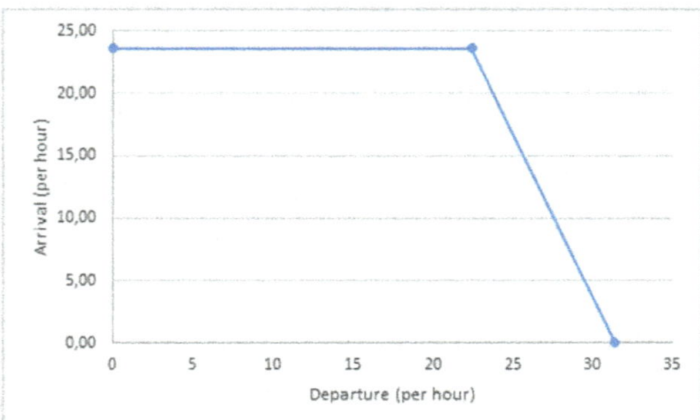

Fig. 1. Hourly runway capacity depending on the type of performed operations.

Figure 1 shows clear that the number of arrival operations remains stable, despite the increase in the number of departures. It means that gaps between arrivals occurred and were filled in. However, when the limit value is reached, an increase in the number of departures results in a decrease in the amount of possible arrivals per hour. The maximum runway capacity for departures is achieved when there are no arrivals during the analyzed period.

3 Simulation Model

3.1 Model Basics

Aimsun Next software was created to model small and large transport networks. It can cover from one single intersection to an entire region. This software is used to create models of intersections, highways, cities, as well as to simulate different types of traffic together with its adaptation to existing transport conditions [15].

As it was mentioned above, within the scope of this research, the Aimsun Next software is used to make air traffic simulations on the runway. The simulation was performed based on the calculation for mixed operations, according to which the selected runway permits 23 arrival and 22 departure operations. Based on this information, as well as on Table 2, a flight schedule was prepared as input data for the model (Table 3).

Table 3. Flight schedule used to create a simulation model.

	arrival operations				departure operations			
	B	D	E	F	B	D	E	F
1	00:00	06:36	12:21	03:47	07:46	13:21	04:37	01:45
2		14:59	17:39	09:32		18:39	10:22	50:03
3		20:17	22:57			21:27	16:09	
4		25:35	28:15			23:57	45:07	
5		30:53	33:33			26:45	47:35	
6		36:11	38:51			29:15	57:27	
7		41:29				32:03		
8		43:57				34:33		
9		46:25				37:21		
10		48:53				39:51		
11		51:21				42:39		
12		53:49				52:31		
13		56:17				54:59		
14		58:45						

Source: own work

It presents categories of aircrafts and times, when they appear in the simulation. ROT for each type of aircraft was also preserved. Calculated separation minima were also taken into account.

This flight schedule was created to perform a simulation with one-hour duration. It was done in view of the above calculation results, which concern the maximal runway capacity only for 60 min.

To summarize, the flight departures and arrivals were introduced to the Aimsun Next package using a specially created flight schedule. To simplify the modelling process, the software function which applies to create a public transport plan was used. Thereby, Table 1 cells marked with identical color present the type of aircraft which were placed under the same plan: they appear in the simulation using established intervals.

3.2 Assumptions of the Simulation Model

When creating the simulation model, there was a number of challenges related to the adaptation of Aimsun Next software functions. It was done with the aim to present and model air traffic on the runway as close to real requirements and conditions as possible. This sub-chapter discusses the steps taken to create the model as well as the assumptions and limitations associated with this process.

The simulated runway shown in the Fig. 2 has a real length, but its width has been set at 20 m due to the limitations of the simulation software. There was no established impact of the reduced width on the simulation correctness. That is why it was assumed that such restriction has only an aesthetic effect. The number 1 indicated the beginning of the runway - a place where arriving aircraft appear. The ROT of arriving and departing aircraft starts to be measured when the point with the number 2 is crossed. Point number 3 indicates the place where departing aircraft appear in the simulation. After crossing point number 4, the departing aircraft leaves the runway - the measurement of its ROT stops at this point. The arriving aircraft leaves the runway as soon as its entire surface in taxiway reaches point number 5.

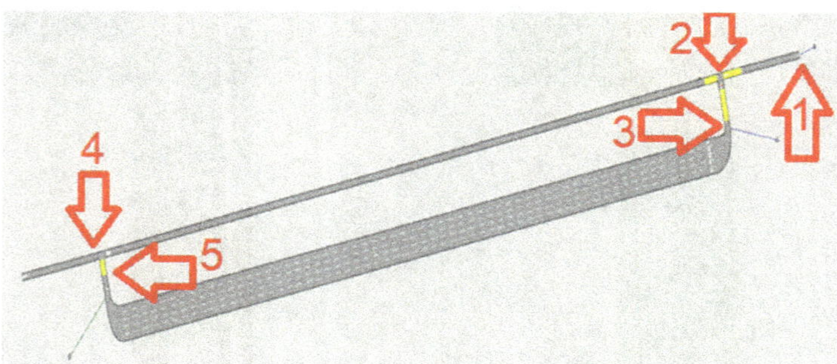

Fig. 2. Runway with taxiways created in the Aimsun Next.

The Fig. 3 shows the exact moment when ROT measurement starts and finishes for arrival and departure operations. Areas of public transport stops which were used by the runway simulation to determine the location of aircraft in a certain place are marked by Aimsun Next yellow in Fig. 1.

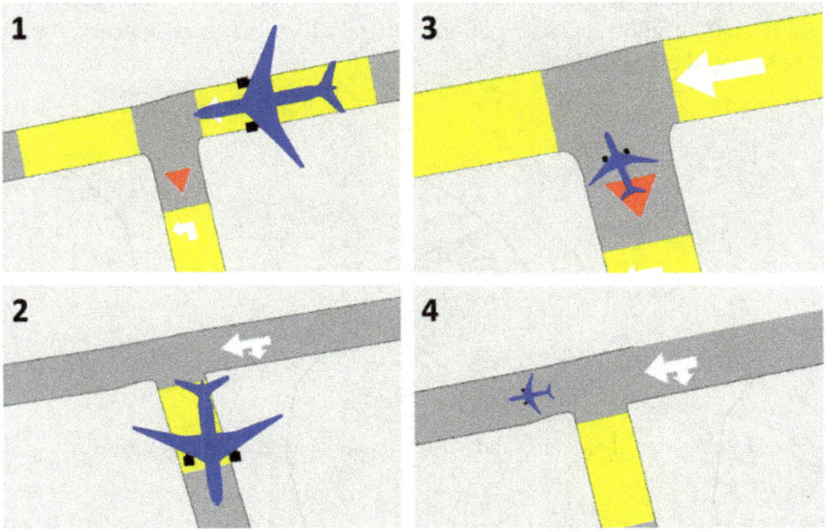

Fig. 3. ROT measurement: 1 – start of ROT measurement for arrival operation; 2 – finish of ROT measurement for arrival operation; 3 – start of ROT measurement for departure operation; 4 – finish of ROT measurement for departure operation;

Speed limits of 300 km/h on the runway and 55 km/h on the taxiway have been taken into consideration. The taxiway was created mainly for illustrative purposes: it helps to facilitate the location of aircraft in the case of a simulation error.

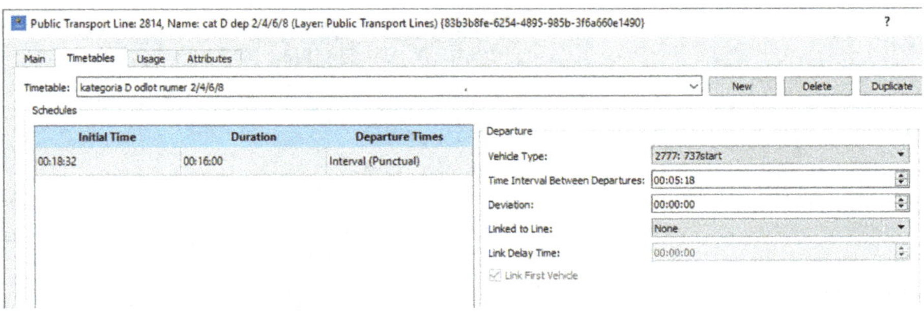

Fig. 4. Usage of a public transport plan to simulate departures number 2, 4, 6, 8 of Category D aircraft.

Figure 4 shows the software function which allows to create a public transport plan. As an example, category D aircraft are considered. The plan starts at 18 min 32 s of the simulation, when a category D aircraft appears on the taxiway at the point with the number 3. The duration of this transport plan was set at 16 min, while the intervals between departures were set every 5 min and 18 s. This means that within 16 min of the simulation there will be 4 departures of category D aircraft.

Figure 5 presents the created aircraft of specific length and width depending on the type of aircraft. It allows to show an aircraft of which category appears at a certain moment during the simulation.

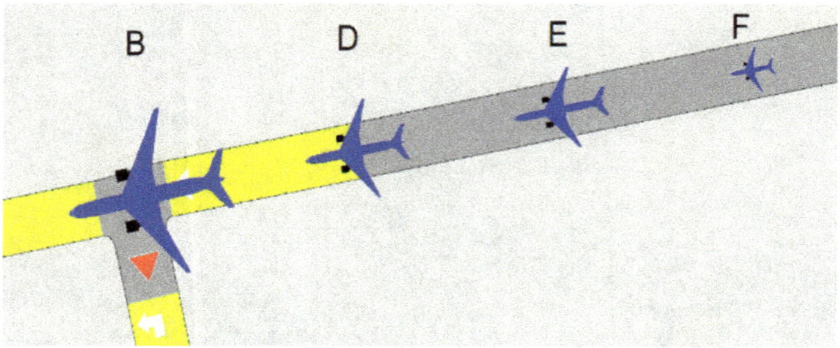

Fig. 5. Size of aircrafts used in the simulation depending on categories.

One of the main challenges during the process of model creation was to simulate the movement of aircraft on the runway taking into account different categories and their main parameters, which are presented in Table 4. Fulfilment of the conditions of both ROT and takeoff speed in the same time seems unrealistic. The last column of the Table 1 presents takeoff speeds reached in Aimsun Next for the simulation, which was oriented to meet the condition of ROT: the worst speed result was shown for B category aircraft.

Table 4. Takeoff speeds of aircrafts depending on their categories.

Aircraft categories	ROT [s]	Takeoff speed		
		Needed [kn]	Needed [km/h]	Reached in Aimsun [km/h]
B	100	165	306	150
D	70	145	269	215
E	60	135	250	245
F	50	130	241	222

Source: own work

The reason for such a failure seems to be very clear: category B aircraft require to reach the highest speed in comparison to other categories, but at the same time they spend the largest amount of time on the runway. Thereby, the maximum speed for category B was set as 275 km/h. In order to maintain the required ROT, it was necessary to reduce the maximum acceleration for this aircraft category. As a result, an aircraft remained

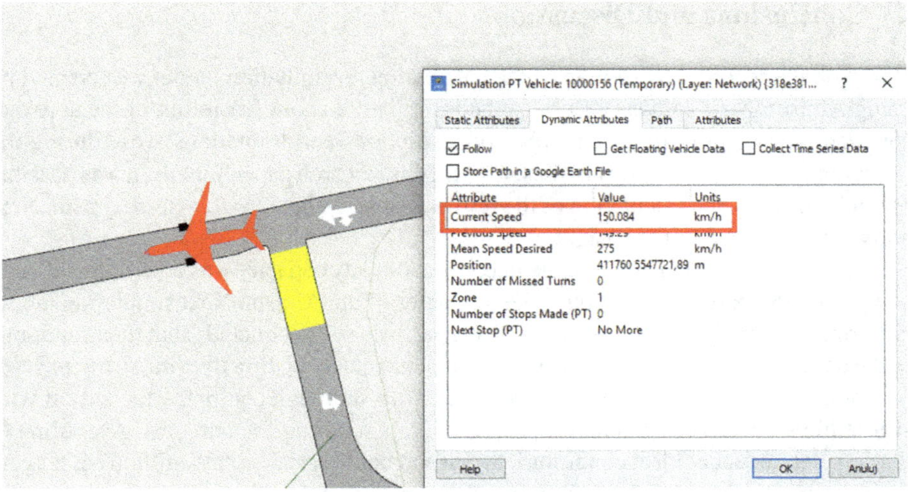

Fig. 6. Takeoff speed reached by category B aircraft.

on the runway during required 100 s. However, the maximum speed reached was only around 150 km/h (Fig. 6) instead of the required 306 km/h.

The parameters used to simulate aircraft of different categories in Aimsun Next are shown in Table 5. These parameters were established by the empirical approach based on the large amount of tests.

Table 5. Simulation parameters of aircrafts depending on their categories.

Aircraft categories	Max speed [km/h]	Minimum speed [km/h]	Max acceleration [m/s^2]	Normal deceleration [m/s^2]	Max deceleration [m/s^2]
B	275	0	0,32	4	6
D	280	80	0,84	4	6
E	278	0	1,35	4	6
F	222	0	4,44	12,98	12,98

Source: own work

To summarize, it could be said that, despite the fact that the Aimsun Next software was intended mainly for modelling and analysis of road traffic in the network, the movement of aircraft simulated in the created model is similar to the actual situation of the functioning runway. The performed simulation, despite the difficulties associated with the limitations of the selected software, was adjusted with the preliminary calculation and, therefore, its correctness as well as the adequacy of flight schedule created on its basis were confirmed. Additionally, the most important assumption, which was to avoid a situation when aircraft ascend simultaneously on the runway, was met.

4 Conclusions and Discussion

Based on the real data as well as on the calculation, a simulation model was created by using Aimsun Next software. Besides the information about feasibility of the software for such type of modelling, a number of restrictions and limitations arose during the modelling process, which are also described in this research paper. Finally, it was possible not only to verify the calculation of the runway capacity, but also to assess Aimsun Next software as a tool for such task.

As a result of the calculation, the maximum runway capacity was determined, which is equal to 45 operations per hour. The outcomes of the Aimsun Next simulation model confirmed the analytical calculations. Thereby, it allows to conclude that the limitations in the development of the Krakow Airport are not connected directly with its runway, but first of all with capacities of its terminal and apron, although equipping the airport with ILS of higher categories would support its operation during fog and mist. According to the calculation, under ideal conditions, the airport could serve over 35 million passengers per year, which would rank it close to top Europe airports.

Unfortunately, the values obtained from the simulation model, e.g. aircraft speeds, are not similar to real ones and could not be used as a source of full-featured data. Taking into account this limitation of Aimsun Next, it is still up for debate whether this tool could be recommended to simulate the runway or does the created model only have an illustrative nature, showing how air traffic looks like at the analyzed airport.

References

1. Subramanian, P.: A simulation study to investigate runway capacity using TAAM. Master theses. Embry-Riddle Aeronautical University, Daytona Beach (2002)
2. Barrer, J., Kuzminski, P., William Swedish, W.: Analyzing the runway capacity of complex airports. In: AIAA 5th Aviation, Technology, Integration, and Operations Conference (ATIO), pp. 1–7. American Institute of Aeronautics and Astronautics, Virginia (2005)
3. Kim, A., Hansen, M.: Validation of runway capacity models. Transp. Res. Rec. J. Transp. Res. Board **2177**, 69–77 (2010)
4. Bubalo, B., Daduna, J.R.: Airport capacity and demand calculations by simulation - the case of Berlin-Brandenburg international airport. Netnomics **12**, 161–181 (2011)
5. Kuzminski, P.: An improved *runway*Simulator—simulation for runway system capacity estimation. In: 2013 Integrated Communications, Navigation and Surveillance Conference (ICNS), Virginia, USA, p. 13612469. IEEE (2013)
6. Zou, X., Cheng, P., Cheng, N.: A simulation model for airport runway capacity estimation. In: 17th International IEEE Conference on Intelligent Transportation Systems (ITSC), Qingdao, China, 14772985. IEEE (2014)
7. Pavlin, S., Žužić, M., Pavičić, S.: Runway occupancy time as element of runway capacity. Promet Traffic Transp. **18**(4), 293–299 (2006)
8. Butler, V.: Increasing Airport Capacity Without Increasing Airport Size. Reason Foundation, Los Angeles (2008)
9. Horonjeff, R., Mckelvey, F.X., Sproule, W.J., Young, S.B.: Planning and Design of Airports, 5th edn. McGraw Hill Education, New York, U.S. (2010)
10. Kolos-Lakatos, T., Hansman, R.J.: The Influence of Runway Occupancy Time and Wake Vortex Separation Requirements on Runway Throughput. MIT International Center for Air Transportation, Cambridge (2013)

11. Zografos, K.G., Stamatopoulos, M.A., Odoni, A.R.: An analytical model for runway system capacity analysis. IFAC Proc. Vol. **30**(8), 29–34 (1997)
12. Safrilah, Putra, J.C.P.: Review study on runway capacity parameters and improvement. In: IOP Conference Series: Materials Science and Engineering, 012108, vol. 209, pp. 1–15 (2017)
13. Grebenšek, A., Pavlin, S.: Calculation of the runway capacity of Ljubljana- Brnik airport. Promet Traffic Traffico **16**, 21–26 (2004)
14. International Civil Aviation Organization: Aircraft Operations, Doc 8168 (OPS/611). Procedures for Air Navigation Services, Flight Procedures, vol. I, 5th edn. ICAO, Montreal (2006).
15. Arkell, C.: Legion for AIMSUN: modelling the interactions between pedestrians, light rail and general traffic. In: AITPM National Traffic and Transport Conference, Melbourne, Australia (2018)

Methods of Resource Modeling of Organizational Objects

Nikolai I. Plotnikov(✉)

Scientific Research Project Civil Aviation Institute «AviaManager», 630078 Novosibirsk, Russian Federation
am@aviam.org

Abstract. The development of resource methodology and soft computing is presented as a new direction of ontological design in this paper. The proposed hypotheses of resource modeling consistency postulate that natural language tools: words, concepts, names, vocabulary, grammar, syntax, used in relations to describe objects, can be transformed into formal models that are suitable for subsequent calculations and automated management. The basis of resource design is the formal resource model, on the basis of which the following elements are formed: complexes of contours, tuples of resources. Thus, the resource method allows us to assign numbers to the properties of objects in weak non-metric scales by shifting the assessment procedures to the area of strong metric scales, which establishes new combinations of applying qualitative and quantitative methods of research and development. A sample of practical resource design and modeling is presented based on the example of the air transport complex. The air transport organizations are chosen based on the applied area: a company, a social group, an individual.

Keywords: Resource · Soft computing · Modeling · Management · Properties · Resource complex · Resource contour

1 Introduction

Properties of objects of fuzzy nature cannot be described and calculated using quantitative methods. Fuzzy modeling in the theory of fuzzy sets (TFS) and in the theory of artificial intelligence (AI) solves the problem of describing the properties of objects, their structure and parameters by applying classical optimization methods and metaheuristic algorithms. The difficulties and problems of applying each of these methods are associated with high computational costs of maintaining fuzzy inference, so "it is necessary to apply several algorithms and heuristics simultaneously" [1]. In the scientific literature, there are claims that computational tasks using TFS methods are redundant and cumbersome. The development of a new approach to the resource methodology (RM) and the use of soft computing (SC) in this paper is considered as a non-identical TFS. This circumstance is the probable reason that the founder of TFS (L. Zadeh, first publication, 1954) proposed the SC concept and term 40 years later (1994) [2].

The concept of "resource" as a category in science and practice is associated with the concepts of "consumption", "use", "application", "distribution", "provision" in material,

© The Author(s), under exclusive license to Springer Nature Switzerland AG 2021
A. Kwasiborska et al. (Eds.): ATE 2020, LNITI, pp. 116–130, 2021.
https://doi.org/10.1007/978-3-030-70924-2_10

natural, industrial, technosphere, and ecological environments. “Resources” is everything that is used by consumers. “Consumers” are any objects that consume resources. The problem is that the category of resources is not identified and there is no humanistic study of it, primarily in logic, philosophy, and philology. The known classifications and types of resources are numerous, diverse, and mutually irreducible. Some resources have ratios. For example, material resources are recalculated through financial resources. However, various classes, especially intangible resources, do not have common grounds.

In this paper, the general foundations of resource theory and methodology of design and management are stated. Sources and differences of the resource approach from analogues are presented. Descriptions of the methods of soft formalization of knowledge required for modeling are given. A concept and derivation of definitions is carried out.

2 Foundations of the Resource Theory and Method

Resource methodology in this paper is presented as an original and independent branch of science. In this paper, definitions are introduced with the symbol (Definition-i).

Definition-1. The field of knowledge of ontological reality design, combined on a resource basis, is understood as a new discipline – resourcology.

The basis of resourcology expected resource representation of realities. The resource complex is understood as transformable sources of energy, information and substances involved in appropriate activities. Resourcology has its own ontological foundations. In this paper, resource methodology is presented as an original and independent field of science.

Resource methodology offers its own bases and sources. The theoretical foundations are formed based on the following content: principles, hypotheses, language, terms, conditions, and procedures. This content allows to generate the output of key definitions. Resource design is based on the concept of a resource model (RM). Resource modeling can be defined as displaying object property parameters from a specified definition area to the area of indicator values. Object elements are linked by relationships, the content of which is set in the resource modeling language (RML) procedures.

Resourcology, according to the author, is partially related to the ideas of tectology, systemology: Klir J [3], logic Zinoviev A.A. [4], pseudophysical logic Pospelov Д.A. [5], JSM-method [6], theory of measurements: Pfanzagle J. [7], Fridman A.E. [8], Berka K. [9], Suppes P., Zines J. [10], Ivin A.A. [11], Poya D. [12], Ruzavin G.I. [13], theory of possibilities Dubois D., Prade A. [14], methods of fuzzy sets and soft computations Khodashinsky I.A. [1, 15].

Each of these theories has a different age and degree of formalization; in this work it gives its own elements of knowledge. However, the analysis of the differences between the presented theory, the created sample, the product and the method, in essence, has no basis for comparison with analogs. Presented below is a block diagram of the resource method of design and modeling, Fig. 1.

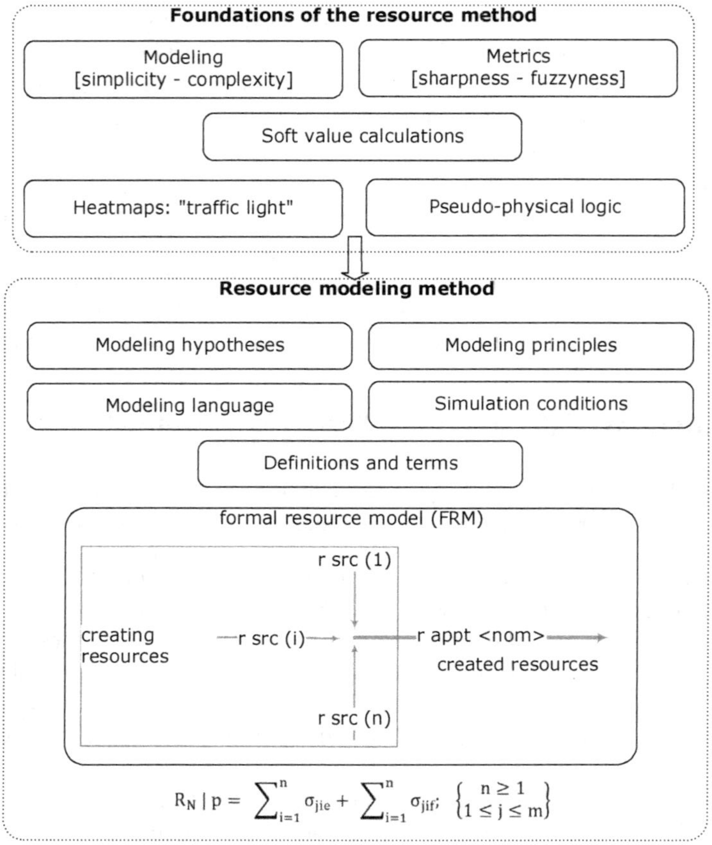

Fig. 1. Resource method: fragment of ontological design

3 Justification of the Method

3.1 The Principles of Resource Modeling

The developed method establishes the following principles of object design and modeling.

The principle of adequacy of observation. The principles prescribe the development of a set of methods for SC object properties. Following this principle, a specific set and methods for observing properties and their changes in states are developed for each object.

The principle of reduction (convolution). The "curse of dimensionality" is related to the complexity of calculating the properties of an object, which increases with the increasing dimensionality of the problem. The solution to the problem is to reduce the dimension of the space to a subspace of a smaller dimension, highlighting the main components for the selection of features using a certain estimation algorithm.

An analogy is the multidimensional space Sn (n-dimensional space $n \geq 4$) as a metaphysical abstraction of three-dimensional space. Mathematical models describing an object can be approximated using geometric images: lines, surfaces, and volumes of three-dimensional space. This is justified by the fact that sets preserve essential properties similar to Sn ($n \geq 4$) similar to the properties of Sn ($n \leq 3$), the study and practice of which already has a huge experience. Therefore, in mental activity, we often use a sheet of paper and draw diagrams in the form of lines, surfaces, volumes. This is how the multidimensional dimension is reduced.

In the combinatorial method of Leibniz Gottfried Wilhelm (1646–1716), everything complex can be reduced to simple things, which are the first principles, the alphabet, and constitute a reduction of exact knowledge. The method is consistent with the laws of hyperbolic "non-Gaussian" distributions. Combinations of first principles, in a fuzzy inventive logic, lead to an increment of knowledge. Leibniz's principles prescribe the empirical and heuristic allocation by the researcher of only essential objects of the subject area that are considered sufficient to describe the object. The informal statement of the reduction problem in object design is formulated as follows: to determine the maximum-minimum number of the most important components that are considered sufficient for an adequate description of the subject area and corresponding to experimental and practical verification.

Example: if the only parameter known about a vehicle is linear movement, movement on a plane or in three-dimensional space, then the purpose (name) of the object and parameters (mass, speed) are estimated with acceptable accuracy. For linear movement, it is a train, for movement on a plane it is a car, for three-dimensional movement it is an aircraft.

The principle of external influence. The principle is based on the following statements: a) it is impossible to design an object only with internal resources; b) it is necessary to use external resources for the design and management of objects. The necessity of external influence agrees with Kurt Friedrich Gödel's incompleteness theorems (1931): "the consistency of a sufficiently rich theory cannot be established by means of the theory itself". In other words, especially large object changes (reforms), require a sufficiently complex infrastructure of scientific, educational, technological, social, and information environment, which can be used to reorganize and reform the object.

3.2 Target Function of Resource Modeling

Heuristic mapping of properties, parameters, indicators, and the selection of scales and evaluation units is based on knowledge of the subject area of the object, which is recognized as axiomatic. The RM includes fuzzy rules (FR) and cluster analysis (CA) for identifying object properties. The same indicator or parameter can belong to several objects or clusters with different degrees of belonging. Example: modeling of dependability properties, indicators of age, time of employment are the same for an individual and a social group.

Clusters are established by postulating the axiomatics of property indicators: their equivalence, independence, and completeness. This is an example of the most general description of the implementation of the subject's needs in transport of natural or artificial nature. The purpose of the change is set by the need of the subject, its target function.

A resource transport complex (RTC) consists of a subject (S), space movement (SM), transport vehicle (TV), vehicle capacity (VC). The objective function defines a destination resource of RCT, that is a type of motion in space. It will be shown how the requirements for the properties of creating resources change depending on the needs of the subject. Table 1 presents the areas of definitions of the subject and the corresponding SM and TV.

Table 1. The needs of the subject in transport

Election areas	Options	Space movement	Transport vehicle	Carriage capacity
[orderliness of movement, freedom of movement]	A →	railroad	train	compartment
	B→	highways	car	salon
[short distances, long distances]	A →	highways	car	salon
	B→	air transport	aircraft	salon
[transportation, self-movement]]	A →	road	bicycle	seat trunk
	B →	road, forest, field	by foot	clothes, backpack

The Appointment Resource. Objects are assigned resource requirements that are consistent with the purpose of the activity. There is a requirement purpose of the activity called appointment. Thus, the target function of needs in the concepts of the resource modeling language (RML) is formed. Therefore, the formation of the target function in the RM is of fundamental importance for establishing the main category of RM called the appointment of the object. The key concept of the resource method is the appointment resource (AR).

Decision Rule. The assignment of activities is carried out in accordance with the rule. The decision rule is formulated like the descriptions in systemology [3], here it is understood as a set of principles and concepts for forming a resource model. All sets and subsets are called resources and are considered finite. The method is based on pseudo-physical logic (PL) of relations of time, space, causality, other relations and their combination [4]. Generalization of any part of the RM object is based on specific relations. For example, a specific flight of an aircraft can be described as the conclusion of the pseudo-physical logic of space-time relations "here and now". The set of relations is formed by the principles of forming key concepts. How well the principles and concepts are developed and justified can be checked by expertise and experience.

3.3 Resource Modeling Hypotheses

In this paper, we propose approaches similar to those used in artificial intelligence (AI), the JSM method [6], natural language (NL). The JSM method is considered as a set of logical-combinatorial techniques of intellectual analysis based on non-classical logics and rules of plausible reasoning, rules for eliminating data uncertainty. In the theory of situational management, the following set of reality model design is selected: "the first step in the engineering analysis of the reality model is to identify groups in the language vocabulary that carry a certain functional load when describing objects and situations (1):

$$NL = \langle C, N, R, ACT, Q, M, MD, E \rangle, \quad (1)$$

where C is the set of concepts; N – the set of names; R – the set of relationships; ACT set of actions; Q – many quantifiers; M – the set of modifiers; MD – multiple modalities; E – multiple evaluations (assessments)" [4].

In this paper, the RML is formed by accepting hypotheses about the richness of NL and its sufficiency for creating formal content. The assumption that any object of artificial origin is less complex in its content and form is consistent with the theory of creating artificial objects.

Hypothesis 1. To describe the studied subject areas and objects of any complexity, it is sufficient to use NL tools. NL tools - words, concepts, names, vocabulary, grammar, and syntax used in relationships to describe objects can be transformed into formal models that are suitable for subsequent calculations and automated management. This hypothesis is consistent with the conclusions and results of this work.

Hypothesis 2. The main means of NL for structural description of the observed world are established relations of words, concepts and names. Relations are displayed by pseudo physical tools, which allows to assign numbers to observable objects of non-numeric nature.

3.4 Terms of Resource Modeling

The common requirements and terms of the RML are set out in the following content.

1. *Limited description of the subject area.* The content of the description of the selected subject area of activity is compiled within the existing knowledge that allows formalizing these descriptions. For example, recognized theories, methods, acceptable mathematical apparatus.
2. *The description language.* It is necessary to select, compile and use a language and alphabet, based on empirical and heuristic expertise, that allows correct decision-making in various situations.

Example: physical fluctuations (vibrations) in flight are evaluated by the pilot as a situation that has occurred: in one case, it is the "shaking" of the engine, in the other it is "swaying" from the effects of atmospheric turbulence. Consequently, the types and causes of fluctuations are different. The moment of identification of the oscillation is fuzzy and is determined by the pilot heuristically "so it may be", and empirically - "so it was".

3. *Solutions*. Simple solutions are to optimize complex situations. Example: the vagueness of the content of situations in flight is integrated into the set of clear decisions of the pilot in flight. This set is a class of standardized flight operations (SOPs). SOPs have features that are expected to produce results that are no worse than the specified nominal value.
4. *Actions*. The dynamics of relations is carried out in actions and operations: a) imperative actions: an order, an indication; b) actions indicating the processes: climb, descent, reading checklist; actions indicating the state of the object: a serviceable aircraft.
5. *Quantifiers*. Generality quantifier ($\forall$) specifies common values and attributes for objects of the class: for example, the statement "all aircraft are in good condition". Existence quantifier ($\exists$) indicates the existence of the truth of a statement for at least one element of the set.
6. *Modifiers* (m_i) are words that establish signs of objects and procedural states, for example, in the flight of an aircraft: "cargo plane", "rapid descent".
7. *Modalities* (md_i) are words that establish the dimension of possibility, necessity, likelihood, belief, plausibility.
8. Evaluation v_i are words that establish the attributes of quality: "good", "bad", "high", "low".

3.5 Conditions of Resource Modeling

The following resource modeling conditions (C) are created.

C-1. *Generalizability*: concepts of different objects must have the generalizing property of a higher-level object.
Example: the aerospace industry generalizes the concepts: design, construction, production, and operation of aircraft.
C-2. *Quasi-identity*: concepts must be of the same cognitive level, have an acceptable identity, similarity.
Note. Similarity, but not sameness. Two car drivers are the same. The driver of a car and the driver of a truck are similar.
Example: an airline and an airport are quasi-identical; they are objects of the same level of air transport organizations.
C-3. *Difference*: the concepts of objects of the same level must be different in purpose.
Example: an airline and an airport are quasi-identical, but also distinguishable by purpose.
C-4. *Isomorphism*: resources form aggregates and classes. Resources of the same class are isomorphic to each other: $\langle \mathrm{r_i} \leftrightarrow \mathrm{r_{i+1}} \leftrightarrow \mathrm{r_{i+n}} \rangle$.

Example: the complex "driver, car, environment" is necessary for all types of transport. They are similar, similarly arranged, organized, and therefore isomorphic.

3.6 Procedures of Resource Modeling

To comply with the stated modeling conditions, the following procedures have been developed for establishing and controlling $\Pi_i(\pi)$.

Procedure (Π_1). According to the quasi-identity condition, the procedure Π_1, is performed, in which the acceptable identity of the concepts of generating resources used for generating a higher-level resource is established (2):

$$\Pi_1 :\cong \{\pi(r_1 \succ r_2), \pi(r_1 \succ r_3), \pi(r_2 \succ r_3)\}, \tag{2}$$

where by a piecemeal comparison (π), the combinations of resources (r_1, r_2, r_3) are recognized as quasi-identical. For this purpose, thinking techniques can be used, such as reducing different concepts to a single concept of a given cognitive level. Example: an aircraft (A) is also a flying vehicle (FV), is also a machine (M); that is, A: $\cong$ $FV \rightarrow M$.

Procedure (Π_2). By the difference condition, the procedure Π_2, is performed, which sets the differences between concepts in the assignment (nomination) of objects. Here, by means of a piecemeal comparison, it is established whether the constituent resources are different from each other (3):

$$\Pi_2 :\rightarrow \{\pi(r_1 - r_2), \pi(r_1 - r_3), \pi(r_2 - r_3)\} \tag{3}$$

Example: the aircraft is moving in the airspace for its intended purpose. FV, by definition, moves in air and airless space. Therefore, when searching for the identity of the concept of "machine", their distinction is also established throughout: A $\neq$ FV.

Procedure (Π_3). According to the generalizability condition, the procedure Π_3 is performed, in which objects are generalized in a single concept of a higher cognitive (epistemological) level (4):

$$\Pi_3 : \{\pi(r_1), \pi(r_2), \pi(r_3)\} \rightarrow R, \tag{4}$$

where the mapping (π) resources (r_1, r_2, r_3) sets a new resource, termed (R). A comparison of the items called: selection, issuing no statements, taking into account; the establishment of compliance: a) reciprocal, b) unambiguous, c) one-to-one; comparison (attributes or values) of properties of objects [4].

Example: objects move through their destination properties in three-dimensional space (3D). The aircraft object is designed to move in the air environment. The FV object is designed to move in an air and airless environment. The submarine (SM) is designed to move in the three-dimensional space of the water environment. Therefore, under the generalization condition, we perform the following procedure Π_3 of mental design (5):

$$\Pi_3 : \left\{ \begin{array}{c} A(air\ space) \\ FV(air\ space + airless\ space) \\ SM\,(water\ space) \end{array} \right\} \rightarrow 3D \tag{5}$$

The presented content is the basis for the adoption of key definitions of resource modeling.

4 Ontology of Resource Modeling

4.1 Key Definitions

The definitions of resource modeling are introduced.

Definition-2. The Resource {R}. An elementary resource r_i is a material, energy, information property of an object of any nature, that is observed by measurement or evaluation in states (parameters, indicators).

Example. The body has the property of gravity and can be in states of motion and rest.

According to logic requirements, one concept, "resource", can be defined through another concept "property", which has the greatest generalization and is the basic one in the resource methodology. The concept of "property" is a category that cannot be generalized more. One can expand the definition of the resource concept in the following way, Table 2.

Table 2. The definition of resource

Objects	Human (pilot)	Machine (aircraft)	Environment (airport)
Property	Parameters and indicators		
omaterial	height	mass	runway length
oenergy	age	speed	passenger capacity
oinformation	flight hours	range	number of shipments per year

The ratio of these or other indicators is called parameters. For example, the parametric relationship and the corresponding requirements for the ratio of the weight of the aircraft and the required length of the runway are established.

Definition-3. *Resource model* (RM) is any set of resources, and it is displayed heuristically.

Definition-4. Resource level R_L is a set of appointment objects of the resource complex R_M, defined by the $L = \{1, 2, \ldots, \delta, \ldots, l\}$ functional characteristic of the complex organization.

Example: the assignment level (L), objects, and execution organization can be represented as follows, Table 3.

Table 3. Resource organization for assigning TC objects

Level (L)	Objects TC	Execution organization
1.Industry	Transport	Ministry of transport
2.Agency	Air transport	Rosaviation, FAA USA
3.Sector	Aviatransportations	Airline
4.Function	Flight operation	Flight complex
5.Composition	Flight execution	Crew
6.Unit	Operation	Pilot

Definition-5. Resource cycle (RC). Resource cycle - processes (stages) of creating an object's destination complex in time.

Example: RC of a technical object: idea → research → development → design → construction (design) → production (manufacturing, construction) → testing (testing) → certification → operation (application, use) → engineering → maintenance → repair → storage → conservation → disposal.

4.2 Formal Resource Model (FRM)

The following definitions are derived from analogs of systemology called directed, neutral, generating, and generated systems [3].

Definition-6. The model that is generated by the design procedures Π_{1-3}, is called the *formal resource model* (FRM).

Definition-7. The model for which the Π_3, procedure can be performed is called the *resolvable formal resource mode*l (RFM).

Definition-8. The set of principles, conditions, and procedures for designing Π_{1-3} is called *resource modeling rules* (RMR).

Definition-9. A resource that is established in compliance with the conditions of quasi-identity (C2) and difference (C3) is called a *generating (creating)* resource. The generating resource is the r_{src} consumed source and specifies the value requirements for creating the appointment resource: $r_{src} :\to r_{appt}$. Example: vehicle (ability to move in space).

Definition-10. A resource that is set in compliance with the generalizability condition (C1) is called a *generated (created)* resource. The generated resource is an appointment resource and calls the requirements of a value corresponding to the named nominal purpose of the activity: $r_{appt}\langle nom\rangle$. Example: energy resource (fuel) for a vehicle.

Creating or generating resources are understood as sources and are called *converging*. A resource that is created or generated is understood as an *outgoing* resource. Outside of this population, resources are not finite and have a possible measure of being generating or generated. The laws of generation can be natural, objective, or artificial, subjective, or heuristic.

Definition-12. An elementary resource contour (RC), (R_N) is a four-part object of any nature, of which three resources are generating, and the fourth is a newly created resource of a higher epistemological level, generated for the target function of activity (6):

$$R_N\{r_{src}(n)\} :\rightarrow r_{appt},\ n \geq 2 \tag{6}$$

where (r_{src}) is the generating resource, (n) is the number of generating resources, and (r_{appt}) is the generated resource being created.

Setting the FRM-fours (3 → 1): "three creating resources → created resource" for forming an appointment resource in this paper is a heuristic choice based on the laws of hyperbolic distributions. It is assumed that the number of generating resources can be from two to five $\langle(2-5) \rightarrow 1\rangle$. This choice of quantity is based on the principle of convolution (reduction) of a set of objects to the necessary minimum for the design of resource complex contours. Example: {transport environment, vehicle, driver}: → transport, Fig. 2.

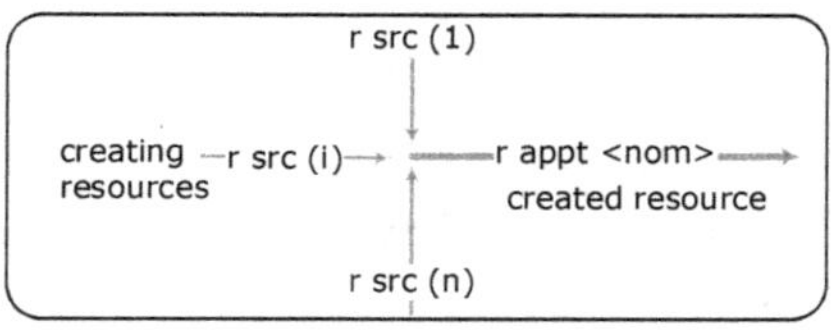

Fig. 2. FRM

Based on the above, we deduce other definitions and postulate the following axioms. Using the rules of formal logic for constructing groups of subjects [4], the following justification for the FRM is introduced. Subject FRM: $\left(r^1, r^2, r^3\right) \rightarrow r^4\right)$ is a subject.

Definition-13. Subject $\left(r^1, \ldots, r^n\right)$ is a unary subject if and only if $r^1, \ldots, r^n$ are subjects.

Axioms:

A-1. $\sim (r^i \rightarrow (r^1, \ldots, r^n)) \cdot \sim (r^1, \ldots, r^n) \rightarrow r^i)$ $(i = 1,\ldots, n;\ n \geq 2)$;
A-2. $(r^i \rightarrow r^k) \rightarrow ((r^1, \ldots, r^n) \rightarrow \rightarrow ((r^1, \ldots, r^n)\ (r^i / r^k))$ $(i = 1,\ldots, n)$.
Otherwise, no other method of forming a FRM, proposed for solving a theoretical or practical problem, is accepted and is possible. The rules for constructing terms included in the unary subject (FRM) are written in the form of the statement: "if (r) is a term,

then (r^i) is a term", or "if the meaning of the term (r) is known, then the meaning of the term (r^i) is known" [4]. In the future, the design of organizational and social objects is postulated by the following axiom.
A-3. The term "individual" (social unit) and the term "class" (social group) are the terms [individual · group], that is, the object denoted by each of the terms whose meaning is known.
Consequence 1. The subjects of FRM terms are investigated, studied and designed in separate meanings, predicates, and sets of predicates in the content and scope of concepts.
Consequence 2. The subjects of FRM terms are disidentified, and their arbitrary replacement by others is unacceptable. Example: the terms "individual" and "group" cannot be arbitrarily replaced, since they are different terms.

In contrast to semiotic models, where generalization is based on the values of object weights, in resource modeling, the contour formation is based on the axiomatics of equivalence of a certain set of resources, which are heuristically established on the pseudo-physical logic of relations. Example: airline, airport, and air navigation organization are compiled in the air transport complex on the axiom of organizational equivalence.

4.3 Domain Model

A fundamentally important step in object design is to establish a basic resource model that corresponds to the target function of assigning the selected subject area. This establishment is carried out heuristically, on the basis of doctrinal (non-reasoned) knowledge of the subject, however, based on the above cited hypotheses, principles, conditions, procedures of RM. For each of the source resources, the following polygons are also designed with a set of elements for the corresponding polygon. It forms a resource complex.

Definition-14. *Resource complex* R_M is a set of objects consisting of m resource contours, the composition of which is set as necessary and sufficient for the design of the subject area of activity.

The resource complex consists of resources from N-contours. The resource complex $R_M|M \geq \{2, \ldots, j, \ldots, m\}$, which is used to create a complex of a higher epistemological level, is called *generating* and is denoted as $\overline{R_M}$, Fig. 3.

A resource complex $R_M|M \geq \{2, \ldots, j, \ldots, m\}$ that is created by other complexes of a lower epistemological level is called a *generated* resource complex and is denoted as $R_{M \succ \overline{M}}$ (7):

$$R_M : R_N\{2, \ldots, j, \ldots, m\},\ m \geq 2 \tag{7}$$

In design, there is a need and possibility of compiling resource aggregates by functional specialization. Then the set of resources forms a tuple of specialization.

Definition-15. *Resource tuple* (R_C) is a set of objects of the resource complex, consisting of k elements m resource contours of a specific functional specialization of activity (8):

$$R_C : R_M(k),\ k \geq 2 \tag{8}$$

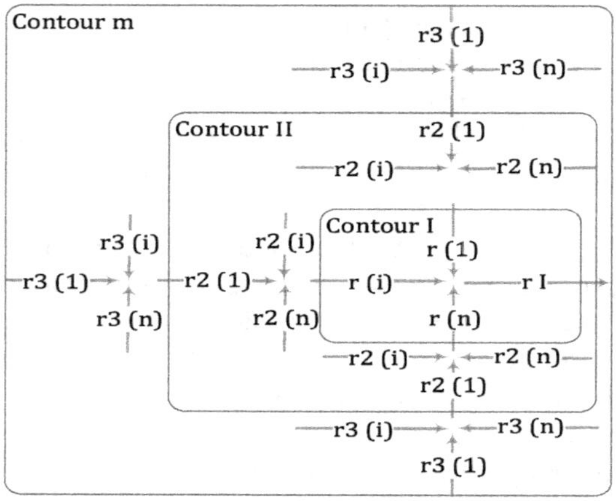

Fig. 3. Resource complex

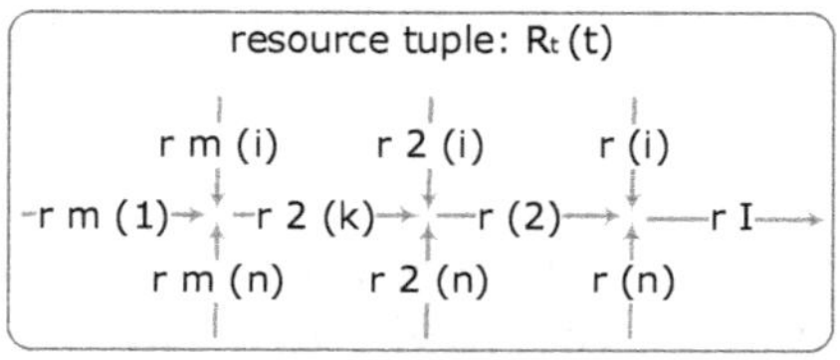

Fig. 4. Resource tuple

Example: {driving school, car, car driver} is a specialty of auto driving. Figure 4.

In the research and design of activities, it is necessary and possible to compile resource aggregates of the entire subject area or part of it.

Definition-16. A model that contains a set of resource descriptions for the (n) contours of (i) levels and (m) tuples is called a FRM of the subject area of the projected object.

The original generalized formula for calculating the values of resources is output as follows. Let the object description, called the resource contour R_N, classify n resources $r_1, r_2, \ldots, r_i, \ldots, r_n$, the property of each of which is defined by 1, 2, …, j,…,m indicators. The property (p) or value of the resource contour $R_N|p$ is calculated as a probability measure in the domain of definition [0, 1] (9):

$$R_N|p = \sum_{i=1}^{n} \sigma_{jie} + \sum_{i=1}^{n} \sigma_{jif}; \left\{ \begin{array}{c} n \geq 1 \\ 1 \leq j \leq m \end{array} \right\} \tag{9}$$

where σ_{jie} и σ_{jif} are the weights of the j-th indicator of the i-th resource, set on a heuristic and formal basis.

Below is a sample of practical resource design and modeling of objects in the subject area based on the example of the transport complex of air transport.

4.4 Practical Design Example

For the transport complex of air transport, the following is to be established: a) flight of the aircraft - the central process of the aerospace industry to meet the need for air transportation; b) the main object of process control is the pilot; c) to make samples of the design of the TC resources and its structural elements with names on NL. In accordance with the terms, procedures, and rules of resource modeling, we will show how the complex of the TC domain is designed: I "flight resources" contours, II "airline resources" contours, III "crew resources" contours, IV "pilot resources"contours, Fig. 5.

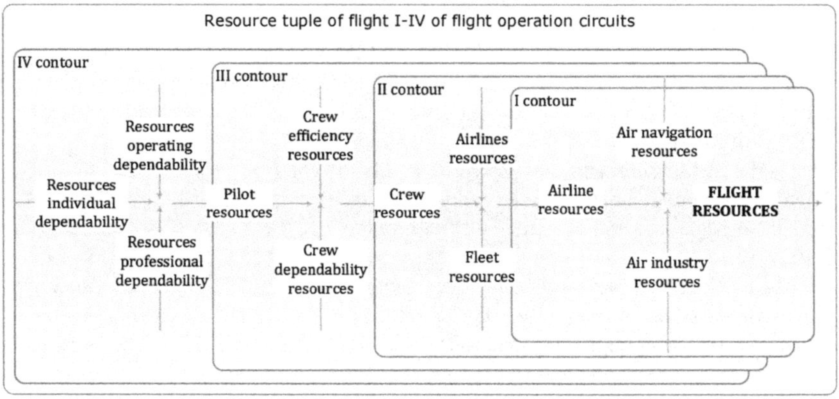

Fig. 5. Practical design example

The full content of ontological design and a example of applied design development are presented in the author's paper [16].

5 Conclusion

An original resource method has been developed, which has technical-economic and natural-scientific bases, a universal theoretical and applied character in modeling and management of organizational social and economic objects. The resource method has the ability to reduce (collapse) the description of properties of objects of any kind, which overcomes the problems of complexity of design, modeling and management. The method contains conditions, rules and procedures for modeling based on the theory of pseudo-physical logic and soft computing. It allows assigning numbers to the properties of objects in weak non-metric scales by shifting the estimation procedures to the area of strong metric scales, which establishes new combinations of applying qualitative and quantitative methods of scientific research and development.

The main provisions of the resource approach create the basis for the development of a scientific discipline-resource methodology (resourcology), which provides new opportunities for designing, modeling and managing objects of any nature. The method allows calculating the properties and states of dependability, risk, and safety, and is the basis for applied research.

References

1. Khodashinsky, I.A., Sinkov, D.S.: Identifikatzya netchetkich system: metody y algoritmy. [Identification of fuzzy systems: methods and algorithms]. Problemy upravlenia – Management Problems, vol. 4, pp. 15–23 (2009). (in Russian)
2. Zadeh, L.A.: « Fuzzy Logic, Neural Networks, and Soft Computing » / Zadeh, L.A // Communications of the ACM, vol. 37, no. 3, pp. 77–84, March 1994
3. Klir, J.: George Architecture of systems problem solving. Radio and Communication, p. 544 (1990). (in Russian)
4. Zinoviev, A.A.: The logic of science, Science Publ., Moscow, p. 279 (1971). (in Russian)
5. Pospelov, Д.А.: Situatzionnoe upravlenie. [Situational management], Science Publ., Moscow, p. 286 (1986). (in Russian)
6. Anshakov, O.M., Fabrikantova, E.F. (eds.): JSM-metod avtomaticheskogo porozdenya gypotez: logitcheskie I epistemologicheskie osnovania. [JSM-method of automatic hypothesis generation: logical and epistemological grounds], LIBROKOM Publ., Moscow p. 432 (2009). (in Russian)
7. Pfanzagle, J.: Measurement theory. World, p. 248 (1976). (in Russian)
8. Friedman, A.: Fundamentals of metrology. Modern course, p. 284, St-Psb. (2008). (in Russian)
9. Berka, K.: Measurement: concepts, theory, problem / K. Berka. Progress, p. 320 (1987). (in Russian)
10. Suppes, P., Zines, J.: Psychological measurement: measurement theory basics. Psychophysical scales (Luse, R.D., Galanter, E.H): Transl. from Engl. World, p. 196 (1967). (in Russian)
11. Ivin, A.A.: Founding logic of evaluations / Ivin, A.A. MGU, p. 229 (1970). (in Russian)
12. Poya, D.: Mathematics and plausible reasoning / Poya, D Science, p. 464 (1975). (in Russian)
13. Ruzavin, G.I.: Probability and plausible reasoning / Ruzavin, G.I // Philosophy of science. – Issue 2, Epistemological and logical- methodological problems, pp. 163–191 (1996). (in Russian)
14. Dubois, D., Prad A.: The theory of possibilities. Applications to knowledge representation in computer science. Transl. from French, Radio and Communications, Moscow p. 288 (1990). (in Russian)
15. Hodashinsky, I. A.: Soft methods of evaluation: monograph / Hodashinsky, I.A – Tomsk: TUSUR, p. 152 (2007). (in Russian)
16. Plotnikov, N.I.: Resucy bezopasnosty transportnych kompleksov. Monographia. [Transport complex safety resources. Monograph], AviaManager Publ., Novosibirsk, Russia, p. 286 (2013). (in Russian)

The Concept of Merging Arrival Flows in PMS for an Example Airport

Magda Roszkowska and Anna Kwasiborska(✉)

Faculty of Transport, Warsaw University of Technology, Koszykowa 75 street, 00-662 Warsaw, Poland

magdaroszkowska0309@gmail.com, anna.kwasiborska@pw.edu.pl

Abstract. Air traffic management must develop to a more comprehensive standard aimed at increasing the complexity of the structure of airspace. It requires implementing new solutions to optimize the efficiency of sectors, as well as modifying current operating methods. A broad overview of airborne sequencing and merging practices which focuses on mathematical formulations and methodologies, also considering different optimization criteria, is provided. Efficient and feasible managing of arrivals involves not only strategic decision-making but also tactical control. In this paper, a novel concept of integrated aircraft trajectory planning in the PMS (Point Merge System) was devised for operations realized to RWY33 (Runway) at EPWA (Warsaw Chopin Airport). The created system is derived from Eurocontrol's guidelines which are based on deployed PMS procedures. Airspace configuration in TMA EPWA (Terminal Maneuvering Area) imposes a specific structure of the designed configuration. It relies on the operational route network indicated in STAR (Standard Terminal Arrival Route). A systematic approach prototype is aimed at dynamically managing the large trajectories in routine dense operation with automated conflict-free solutions. Although such an anticipation introduces early limitations in operational management of upstream sectors, it reduces the need for holding and is beneficial in terms of standardization of working methods and flight efficiency.

Keywords: Point merge system · Air Traffic Management · Airspace capacity

1 Introduction

1.1 Background

The increase in air traffic flow results in aircraft congestion in TMA as one of the most constraining factors which affect the planning and scheduling strategy [1]. The ATC (Air Traffic Control) controllers are responsible for monitoring and controlling the air traffic. In accordance with flight phase, the ATC is divided into ACC (Area Control), APP (Approach Control) and TWR (Aerodrome Control) [12]. The article discusses the TMA which is the area of APP responsibility. APP controllers receive information about flight progress, take into account the performance of aircraft as well as the current operational situation (aircraft altitude, atmospheric conditions, emergency accidents)

© The Author(s), under exclusive license to Springer Nature Switzerland AG 2021
A. Kwasiborska et al. (Eds.): ATE 2020, LNITI, pp. 131–145, 2021.
https://doi.org/10.1007/978-3-030-70924-2_11

and set the final order of aircraft on the runway in use. They keep aircraft vertically or horizontally separated guiding them along predefined routes. APP controllers stay in consistent contact with other ATC services and work in teams of two or three to more precisely split tasks. The process of managing the arrival flows to the airport is divided into three major segments: strategic, pre-tactical and tactical. At the strategic segment, there is developed an initial plan that provides runway capability for landing flights. As far as the pre-tactical segment is concerned, there are collaborative decisions made involving operational stakeholders. The tactical segment includes updating the plan based on current demands. The TMA is considered as the most complicated portions of airspace, as far as the arrival and landing of aircraft is simultaneously operated. Because of the many interacting elements, it has become a complex process to manage the arrival and departure flows to the airport. With the support of adequate tools such as AMAN (Arrival Manager System) or DMAN (Departure Manager System), it is easier to anticipate the metering of constraints and consequently the air traffic flow is coordinated more accurately. The authors [2] present the main features of the indicated systems and explain in detail the performed coordination algorithms relying on fuzzy logic.

1.2 Literature Review

A number of studies present the scheduling process as a field of classical queuing theory. It is considered then as an assignment of arrival units to some machine service. A sequence is formed, or the aircraft joins a queue (unless the service is available at once). When the service is completed, the aircraft leaves the system [15]. The application of broadly defined theoretical algorithms determines the methodology concept in the aspect of minimization of schedule length [14]. However, as computational measurements prove, not all of the applied techniques are effective enough in practice, e.g. the best order in terms of efficiency in the sequencing strategy FCFS (First Come First Served) [5] is rarely indicated, or task planning is processed on machines with limited performance in a deterministic approach [26].

The scheduling theory is concerned with work practices in a simulation model for Rome Fiumicino airport (LIRF) [7]. The mechanism distinguishes both simplicity and universality. The authors point to the prospect of its general implementation at any airport. With the use of a module based on Petri Nets, a decision support tool is made that evaluates landing aircraft sequence [23]. In the scheme of distribution of workflow, there are intentionally introduced frequent interferences in order for the effectiveness of various control strategies to emerge. In addition, with the use of stochastic Petri Nets, the environment which pertains to initial formation of the landing aircraft stream [22] is simulated.

The recent research is focused on the implementation of dynamic programming solutions which postponement more efficient queue formation [18, 19]. The presented results indicate that the system affects the delay absorption capacity and improves the processing time. On the other hand, the paper [6] concentrates on solving the issue of aircraft sequencing by using a genetic solution. The deliberations also involve a specific system of management of air traffic flow based on the division and cut algorithm [25]. The system handles a set of separately realized n tasks (either take-offs or landings) related to one machine (runway). Introducing some modifications to the created mechanism may be used as a subject of further analysis for more complex models, including different aircraft weight classes or airports with more runways. In the paper [25], a multi-objective model is developed which contains a fuzzy programming application for the sequencing of aircraft landings. The mechanism implements the queue and indicates deviations from the estimated landing time. It is stated that the system reflects the actual operating conditions more accurately in comparison to deterministic simulation models.

The landing time generates costs according to the time it takes an aircraft to descend, and to the delays which are associated with it. Therefore, the most common used criteria for the optimal scheduling method is to minimize the average time of the service process. The conducted study [13] has shown that reduction of merge points shortens the execution time for landings. There are also considerations [24] which determine the scheduling strategy in which landing time is related to airline preferences. Furthermore, there are set ideas which provide different optimization goals e.g. The "Just-In-Time" concept. It is a multi-criteria performance aimed at leveling the workload balance depending on the number of passengers, landings per airline and number of landed passengers per airline over the planning horizon [5]. The paper [16] presents a mixed-integer linear programming (MILP) approach to optimize the sequence of landing aircraft. The proposed model has been applied, through numerical simulations, which revealed the best performance in terms of delay reduction and air traffic controllers' intervention rate. Such an application might play the role of a supporting tool which better coordinates with terminal maneuvering area operations. Moreover, the proposed approach strategy can be conducted in areas with a high flight density.

A comparative study [27] provides an automatic generation system of Curved and Continuous Descent Approach (CCDA) trajectories. The issue of sequencing and merging techniques is considered in the overview of CDA-PM integrated methodology. Potential benefits were investigated regarding fuel consumption, capacity as well as environmental impact.

The other paper [17] focuses on the approach attitude, based on Multi-layer Point Merge (ML-PM) route network. With a multiobjective program, which combines both the PMS benefits and automated separation assurance, conflict-free as well as more efficient arrival trajectories are generated. Moreover, authors describe other potential operational aspects in terms of flight efficiency, like average landing interval, delay and runway throughput. Even though a number of techniques for scheduling aircraft have been developed, as air traffic grows, so does the need for efficient, globally harmonized and interoperable ATM (Air Traffic Management). A performance-driven and advanced data-processing technology enhances the ATM framework and ensures the sustainability of the aviation sector in Europe. That is why the SESAR (Single European Sky ATM Research) project [21] was launched, implementing the PMS fits in SESAR targets and future operational improvements [7]. The authors [4] conduct a model-based simulation of the PMS. Validation shows that the method disposes the airspace capacity sufficiently and enables an effective sequencing even under high air traffic congestion. To accomplish it, an operating method [10] was elaborated. Further investigations performed in this paper indicate design keys for new concept of merging arrival flows in the PMS to the EPWA.

Regarding the prosecuted literature review, PMS has a substantial potential to improve and harmonize arrival operations in terminal airspace. Nevertheless, it is necessary to conduct further stimulations to confirm the full applicability of the system considering all supportive arrival integration systems. The article details a practical sequencing and scheduling algorithm for the system designed for the EPWA that may substitute current operational method, which is included in STAR.

1.3 The Main Principles

The PMS is a new technique for integrating arrival air traffic flows to the airport in areas under high air traffic load. It was conceived at the EUROCONTROL Experimental Centre in 2006 [8]. The new procedure has been implemented so far at twenty-five airports in the world and this number is likely to keep growing rapidly in the coming years. The PMS supports the improvement and standardization of arrival operations, from descent to the final approach. It also harmonizes arrival operations in terminal airspace with a pan-European perspective in conjunction with existing technology, including [4]:

- advanced continuous descent approaches (towards 3D) foreseen from the vertical level stated at the sequencing legs to merge points altitude;
- use of pre-defined RNAV routes (ultimately allocation thereof) with advanced ground support tools;
- airborne Separation Assistance Systems – sequencing and merging (ASAS S&M);
- trajectory-based operations (towards 4D trajectory management, including adherence to an agreed or constrained time of arrival).

The joint use of P-RNAV, PMS, CDA and other supporting tools e.g. AMAN (Arrival Manager) is expected to maximize the efficiency of TMA operations and thereby provides economic, operational and environmental benefits. The PMS aims at creating a structured and intuitive way of building and maintaining sequence, with no requirement for new tools. The indicated procedure reduces the need for controller tactical intervention, by keeping an aircraft on lateral navigation. The system relies on simple closed loop clearances. Instructions are formed mainly for speed and level. Open-loop radar vectors are used to recover from unexpected situations. The concept of merging arrival flows in the PMS must be flexible enough to maintain capacity during periods of high traffic load. It shall provide a built-in capability of delaying or expediting aircraft (through path stretching or shortening) in a more flexible manner than current P-RNAV applications in TMA. The flow of traffic is more orderly sequenced with a contained and predefined dispersion of trajectories. In that light, in the PMS, it is easier to optimize flight routes, maintain internal separations and consequently prevent the risk of a potential collision. The indicated factors improve controller anticipation, situation awareness, division of tasks and cause less ATC workload [8, 9]. The simulations confirm that the PMS makes it possible to handle more traffic safely and efficiently. There is a 40% traffic increase compared to current traffic levels in TMA EIDW (Dublin Airport) [7]. Furthermore, after the introduction of the PMS at the LIRF (Rome Fiumicino Airport), the terminal airspace capacity increased to 50 aircraft per hour [7].

1.4 Design Components

Considering an ordinary configuration of PMS, it is a structure contained in a triangle-shaped area. It consists of (Fig. 1) [4]:

– a merge point - used for traffic integration,
– sequencing legs - path stretching or shortening for each inbound flow.

As for any procedure involving RNAV-flown turns flight patterns, flight paths with fly-over waypoints are provided except for the last point at the end of the sequencing leg where fly-by waypoints are implemented. A pre-defined set of distance markers centered on the merge point is displayed. They allow the ATC to adequately estimate the time of issuing the direct-to instruction.

In practice, the PMS design exhibits an overall symmetry for the sake of remaining the most intuitive operating method. The sequencing legs shall be approximately equidistant which means that they are feasibly at the same distance from the merge point. Furthermore, the sequencing legs may be segmented to provide the iso-distance property which means that distance to the common point shall remain the same all along the sequencing leg. Vertical separation between legs can be achieved through either levelling off or a constrained descent.

Figure 1 presents a PMS configuration with full overlap and level off. Nevertheless, the PMS may be also supported by different route options which are depicted in Fig. 2.

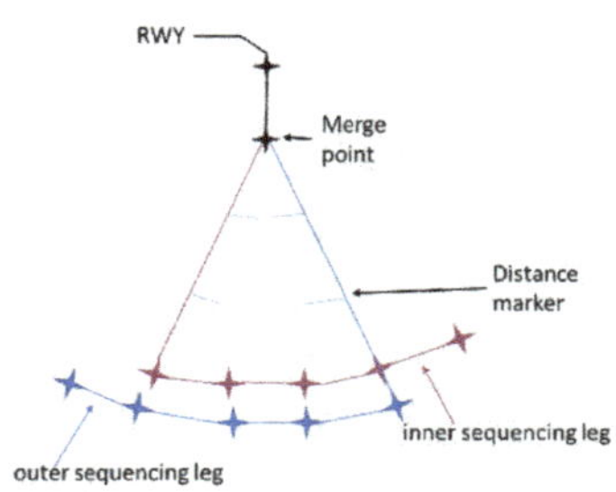

Fig. 1. Example of the PMS with two inbound flows

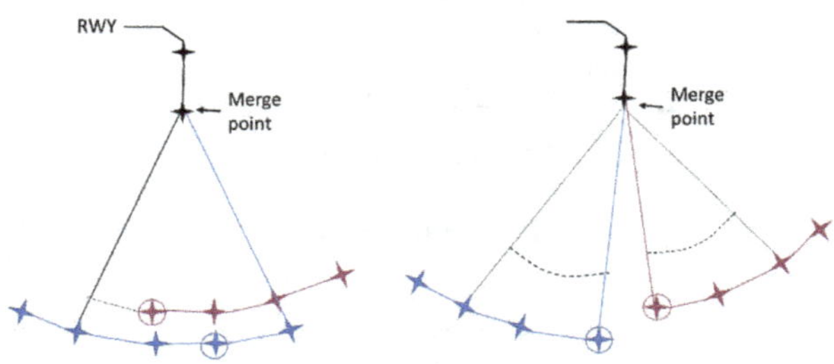

Fig. 2. Possible configurations of sequencing legs

On the left side, a combination of the parallel legs with partial overlap is presented, with constrained descent for the overlapping part (level off is also possible). On the right side, a structure of dissociated non-parallel legs is shown, with an unconstrained descent (a vertical separation may be required between the ends). The choice of an appropriate configuration depends on local constraints and capacity.

The length of the sequencing legs shall be at least 20 NM in order to enable absorbing more delay [4]. The vertical separation should be introduced between sequencing legs with the distance of 5 NM [10]. An adapted buffer along a track prevents the risk of longitudinal separation infringement. Difference in altitude between the sequencing legs cannot be too large and it shall be equal to at least 1000 ft. As a general principle, the inner leg is designed at a higher altitude than the outer one. This is to keep an aircraft at compatible speed for the sequence building and its maintenance. That is why the ATC is provided with more time to react in case of unexpected descending from a higher sequencing leg, or in case of metering a problem at the entry of the system. Furthermore, the spare level is used to vertically separate flights exiting from the same hold located ahead of the sequencing leg entry.

In particular, the symmetry principle impacts both the distances and angles applied in the system. The angles at the following waypoints of the sequencing legs are identified below in Fig. 3 [4]:

- α - the track angle change at the first waypoint on the sequencing leg;
- β - the track angle change corresponding to the direct-to instruction towards the merge point;
- δ - the angle formed by the envelope of possible routes to the merge point;
- γ - the track angle change at the merge point.

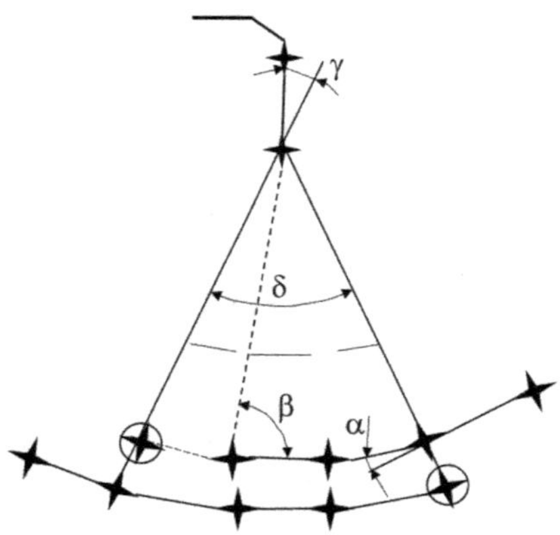

Fig. 3. Angles in PMS

The track angle α at each waypoint is constant. The angle β is almost constant at every point on the sequencing legs and approximately equals to 90°. With increasing the angle δ, the length of the sequencing leg is proportionally greater. When the angle δ is equal to 180° and two aircraft are coming from opposing entry points, there is the potential risk of a head-on situation. As far as the angle δ it is matched, the length of sequencing leg gains at least 20 NM.

Table 1 identifies different sets and parameters for PMS structures which have already been implemented in Europe.

Table 1. PMS at other airports.

Airport name	Initial height [ft]	Final height [ft]	Distance [NM]	Descent rate (ft/NM)
Theoretical PMS	12000	6000	20	300.00
EHAM	5500	2000	13.5	259.26
EIDW	8000	3000	19	263.16
ENGM	10000	5000	18.1	276.24
EGLC	10000	6000	15	266.67

2 Implementing the PMS at the EPWA

2.1 The Terminal Airspace Design

The PMS is expected to be limited by the environmental constraints which impose its specific structure. The system is adjusted locally depending on the airspace sectorization and boundaries in the TMA EPWA (scheme) that are illustrated in Fig. 4.

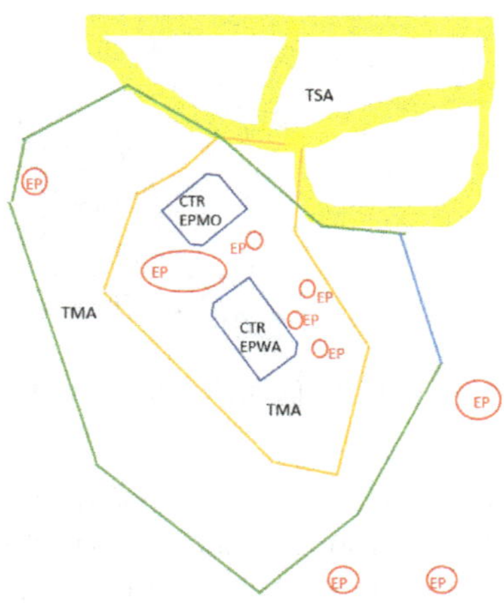

Fig. 4. The internal structure of TMA EPWA

The TMA airspace is used for aircraft landing at the EPWA and the EPMO (Warsaw Modlin Airport). Air traffic flow to EPLL (Lodz-Lublinek Airport) is also organized. The EPRA (Warsaw Radom Airport) is also going to be included, which will distinctly affect the airspace management at the TMA EPWA. The analyzed airspace includes areas where an aeronautical activity is prohibited (those zones are marked red in Fig. 4). In the MATZ

(Military Aerodrome Traffic Zone), the Polish Air Force conducts its activity. The tactical exercises for the Polish Air Force are performed in a TSA 02 (Temporary Segregated Area). West from the TMA EPWA, the MTMA EPLK (Military Terminal Control Area of Lask Airport) is indicated. In TSA 04D, there are mobilized flight trainings from the EPDE (Base of Polish Air Force University). In the region, the ATZ EPBC (Aerodrome Traffic Zone of the Warsaw Babice Airport) is also established. That zone is reserved for polish aeroclub stakeholders. Additional areas like TRA 45 (Temporary Reserved Area) and ATZ EPNC (Landing Strip in Chrcynno) are also provided, and there are organized parachute, paraglider and microlight pilot jumps. Using each of them requires a prior authorization from the ATC.

In accordance with the terminal airspace design, a proposed PMS concept for the EPWA is presented in Fig. 5.

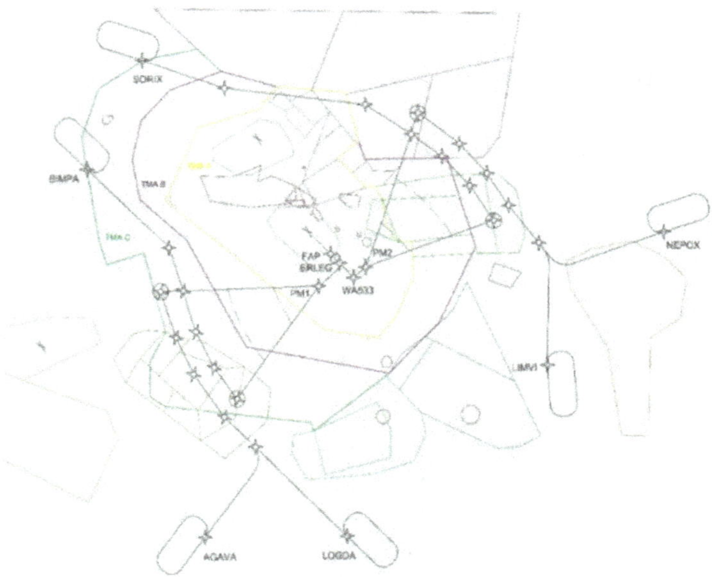

Fig. 5. PMS configuration in TMA EPWA

The stated structure is combined with a symmetrical configuration of two PMS (western and eastern triangles) consisting of inbound flows feeding the appointed runway at the EPWA. Most of PMS structures consist of just two close parallel sequencing legs with a levelling-off constraint in the direction of the final approach path. Nevertheless, implementing such a design choice would result in many difficulties in a management process for the analyzed airspace.

In the southern area of the TMA EPWA there are set prohibited areas - EP P1, EP P28, EP P3. Air traffic flow is going to be guided to the EPRA. Furthermore, there is a large congestion of road crossings. With implementing two PMS combinations, it is expected to achieve an appropriate spacing between two successive flights as soon as the aircraft leaves the sequencing legs. The presented PMS route structure shall enable de-confliction of arrivals flows which significantly increases the airspace complexity as well as absorbs delays. On the other hand, the proposed configuration exposes aircraft to perturbations (wind effect and other weather conditions may vary in each segment of the PMS structure) and restricts the optimization of the initial segments feeding the sequencing legs.

Local constraints limit the dimensions of the implemented system. The western PMS triangle is confined by TMA EPLL. What is more, its flight paths interfere with the MATZ EPTM (Tomaszow Mazowiecki Airport) but the vertical restrictions do not affect the flow coordination movement. The position of the eastern PMS triangle is determined by the vertical altitudes of the MATZ EPMM (Minsk Mazowiecki Airport) as well as the areas where aeronautical activity is prohibited - EP P10, EP P18, EP P9, EP D36. The created concept should also be reconsidered with strategic contribution of flow movement as well as inter-aircraft spacing in mind. It is necessary to smoothly acquire the sequence order from both triangles. That is why the reorganization of TMA through the extension of its boundaries is envisaged.

2.2 The PMS Concept for the EPWA

Figure 6 represents the multiple PMS concept for EPWA, where RWY33 is used for landings. The system's configuration is based on Eurocontrol's guidelines [9, 10, 20, 28]. The performed analysis was supported by current sophisticated ATS requirements outlined in the STAR procedure.

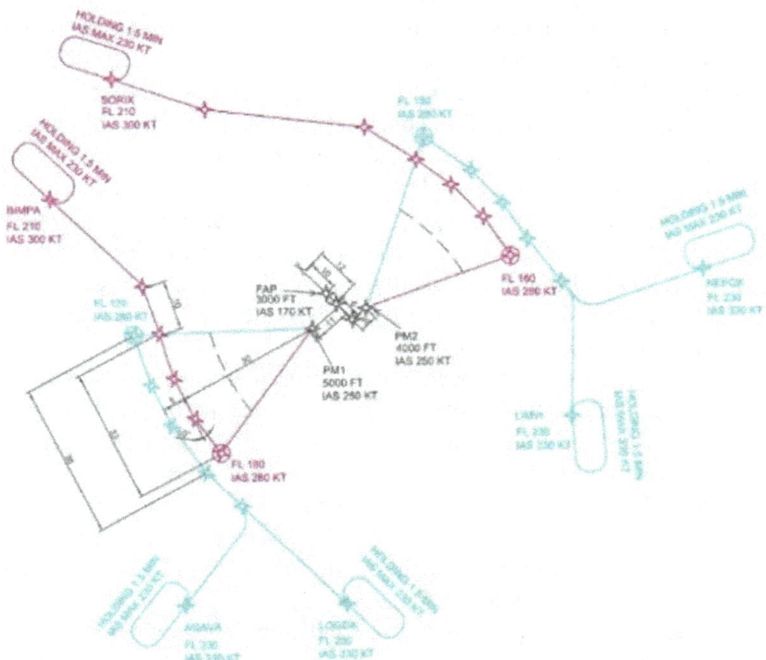

Fig. 6. PMS Route Network Design for EPWA

In comparison to other PMS structures, the created flight path pattern is quite large in size. The sequence order is determined earlier, which triggers the increased predictability of flown trajectories. In the authors' view, such a design option assures more equitable arrival flows management as well better complex in sequence building tasks due to more flexibility in varying altitude and speed values.

The entrance to PMS is provided with six entry points. The western stream flows are led to BIMPA, AGAVA, LOGDA and the eastern stream flows adequately to SORIX, NEPOX, LIMVI. The applied configuration supplies balanced air traffic flow contribution from all directions at the TMA EPWA. Additional entry points would result in a greater dispersion of aircraft. There are situated holding stacks before each upstream which serve as the tool supporting balancing the air traffic. At holding stacks an aircraft flies for 1,5 min at holding speed at or below 230 kt.

The entry points' altitudes are diversified and depend on the distance to the airport. Entry points - BIMPA and SORIX are within the current TMA borders and for these airways it is required to maintain a flight level of 210 FL (as it is referred in STAR). In order to ensure an appropriate separation, the speed restriction is limited to 300 kt. The entry points - AGAVA, LOGDA, NEPOX, LIMVI are not defined in the current TMA borders. Consequently, the defined flight level and the airspeed are increased to relevant volumes - FL 230 and 330 kt.

The considered PMS configuration consists of six inbound flows on two parallel sequencing legs. With more than two parallel sequencing legs, the decision time of issuing the direct-to instruction for the leading aircraft is significantly prolonged. It contributes to the extension of the distance from the sequencing legs to a common point in order to facilitate the adequate flow movement. In addition, such a configuration complies with the overload of spare levels in total. It results in heterogeneous traffic conditions in terms of optimization of vertical profiles for descending aircraft. The equidistant property with 5 NM lateral distance between sequencing legs is observed in both systems. Sequencing legs are designed in combination of full over-lap with level-off. The pre-defined length of sequencing legs defines the maximum path stretching capability, accordingly: 33 NM (the inner leg) and 36 NM (the external leg). The sequencing legs are segmented and they are forming quasi-arcs which are centered on the merge point. Concerning the geometry of the route which joins the sequencing leg, subsequent track angles at waypoints along the sequencing legs (α) shall be small enough (deployed 10°) to avoid a wide divergence between route segments. It allows for a clear anticipation and detection of improper ATC behavior.

Further analysis focuses on the area of the approach airspace performance to merge point and its interaction between concept levels. The results of that consolidation are reflected through diversification of the vertical restrictions on the sequencing legs which depend on the flow movement: eastern triangle - FL 160 (inner leg) and FL 150 (outer leg) and western triangle - FL 180 (inner leg) and FL 170 (outer leg). Indicating higher altitude enables better CDA performance. Nevertheless, the CDA profile is calculated for each different aircraft type and therefore the duration of descending is distinguished. The defined speed limit - 280 kt is equivalent on each sequencing leg to standard operational conditions in the system. The published speed and altitude reductions shall be obtained prior to entering the sequencing legs in order to absorb more delay with the same leg length, as well as to minimize ACAS (Airborne Collision Avoidance System) nuisance alert. It is obligatory to keep the aircraft at a specific value during flight along the sequencing legs.

The vertical separation is defined when the aircraft is about to pass the merge point in descent - the eastern triangle is associated with PM1 - 5000 feet, and the western one is associated with PM2 - 4000 feet. When multiple merge point systems are combined, a lateral offset between merge points must be provided. An applied altitude difference prevents a "nose-to-nose" collision for the flows converging towards the common point. Therefore, the western PMS was placed 6 NM above the eastern one. It facilitates the visual assessment of spacing between two successive aircraft having the same distance to go to the exit of the system at the FAP. A homogeneous flow must be ensured during flight along a common path and allow the standard ILS interception in approach. That is why the provided speed and altitude volumes are likely to be followed by STAR guidelines.

2.3 Aircraft Scheduling in the PMS

Aircraft scheduling in the PMS model relies on a systematic sequence of actions. It includes merging to a common point and path stretching along the sequencing legs. The proposed concept of merging arrival flows may be regarded as a queue modeling process which is accomplished by analyzing the cumulative passage of aircraft as a function of time. The queuing system defines waiting aircraft on the approach $P = \{P_1, P_2, \ldots, P_n\}$. They are continuously distributed into available servers $M = \{M_1, M_2, \ldots, M_n\}$ which are represented by the waypoints. Several subtasks$T = \{T_1, T_2, \ldots, T_n\}$, which must be fully completed by distinct parallel servers, are allocated. The computation time with a server depends on speed and altitude variables. The process is considered from the point an aircraft joins the queuing system to the exit point. Processing time is estimated independently for each aircraft (it results from flow-equivalent conditions of the queuing system). In order to establish state transition rules, the preemptive repeat-different method of aircraft scheduling is assumed, depicted in Fig. 7.

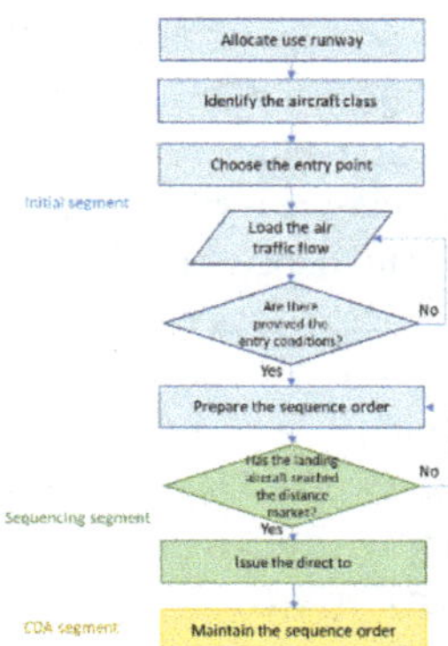

Fig. 7. Flow chart of aircraft scheduling in the PMS

The operating method identifies three segments – the initial segment, the sequencing leg run-off route segment and the CDA segment which are presented in Fig. 8. Each segment requires different performance calculations.

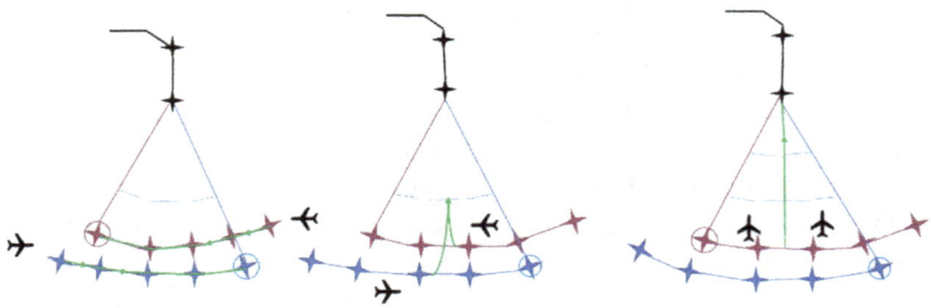

Fig. 8. PMS segments: initial (left), sequencing leg run-off route segment (middle), CDA segment (right).

The initial segment is between the entry point of the PMS to the first point on the sequencing leg. In the first step the P_n is assigned to a runway. Then, the aircraft is classified into one of three weight classes in order to determine appropriate spacing between P_n. Aircraft are served one after another depending on arrival time which makes all the controllers more aware of the sequence. The next step is checking the entry conditions (altitude, speed, separation) to reach the separation requirement within each arrival flow. Spacing could be affected by a wake vortex, weather conditions or air traffic flow. In view of this, the T_n is changed by speed control or issuance of a holding task. The minimum longitudinal separation distance is based on the ICAO vortex separation distances and it is equal to 7 NM within the triangle. The P_n must be metered before entering the sequencing leg.

The segment of sequencing leg run-off route is between the first point of the sequencing legs and the point where the turn to merge point is issued. The P_n receives the direct-to instruction when the P_{n-1} exceeds the next pre-defined distance ring. The processing time of the T_n is variable because the P_n is allowed to turn from the M_n on the sequencing leg at any time. Nevertheless, the prompt issuance of the direct-to instruction is pivotal in performance aspects of the merge point technique. In order to facilitate adherence to entry conditions (traffic deconfliction and descent), it is recommended to release the T_n instruction on the first M_n on the sequencing leg. Unless both the sequence order and an initial inner aircraft spacing are established, the T_n instruction may be issued on further M_n which includes the sequencing leg. It is necessary not to instruct too many T_n towards the merge point. That is why just one P_n should be scheduled for the sequencing leg. If the leading P_n is on M_n of the inner leg and the trailing P_n is on M_n of the outer leg, higher priority is assigned to the P_n with expected shorter service time. It means that it is preferable to serve the operation from the inner sequencing leg first (provided that further path stretching is enabled). In case the P_n reaches the M_n which is at the end of the sequencing leg, the T_{n+1} is automatically assigned without receiving a direct-to clearance. Subsequently a spacing adjustment solely relies on the speed control.

The CDA segment is the final segment where the aircraft performs the CDA towards the FAP. The sequence of the P_n is built and there it is required to keep the inter-aircraft spacing. The buffer in spacing shall be provided with a feasible speed range including 3 NM separation at touchdown [29].

3 Conclusions

The PMS specifies minimum inter-arrival spacing restrictions, precedence in sequence order that prevents overtaking, and time windows. A sophisticated traffic flow model for scheduling aircraft in the PMS is consequently considered. Applied methods of computing the aircraft schedule and landing times are necessary to accommodate practical issues that affect the formation of the sequence.

This article presents the arrival PMS concept procedure to handle the inbound traffic to the EPWA. It is adjusted locally depending on airspace configuration and air traffic distribution. Moreover, the contributions to the queuing process of landing aircraft are considered. The sequence order is subject to the local operational constraints, precedence, separation requirement and possible arrival ways. Nevertheless, it is determined earlier, which increases the predictability of flown trajectories. The presented methods of aircraft scheduling are applied to maximize runway throughput as well as to minimize the processing operational time providing a safe, expeditious and orderly flow of air traffic.

The applicability of the presented PMS concept shall be collaboratively defined and refined after conducting a series of flight simulations that consider all relevant airspace, traffic and operational constraints. The stimulated environment must include more comprehensive and realistic operations (two metering points, overflights, departures, wind, high traffic load, the arrival manager and current constraints) in order to evaluate the system's feasibility, benefits and limitations. The results will prove whether there is a need to introduce any changes to the concept or not.

References

1. Andreussi, A., Bianco, L., Ricciardelli, S.: A simulation model for aircraft sequencing in the near terminal area (1981). https://doi.org/10.1016/0377-2217(81)90004-7
2. Bãhme, D., Brucherseifer, R., Christoffels, L.: Coordinated arrival departure management. In: Proceedings of the 7th USA/Europe Air Traffic Management Research and Development Seminar, ATM (2007)
3. Bianco, L., Nicoletti, B., Ricciardelli S.: An algorithm for optimal sequencing of aircraft in the near terminal area. In: Optimization Techniques (2005)
4. Boursier, L., Favennec, B., Hoffman, E., Trzmiel, A., Vergne, F., Zeghal K.: Merging arrival flows without heading instructions, in Proceedings of the 7th USA/Europe ATM Research and Development Seminar, ATM (2007).
5. Boysen, N., Fliedner, M.: Scheduling aircraft landings to balance work-load of ground staff. Comput. Ind. Eng. **60**, 206 (2011). https://doi.org/10.1016/j.cie.2010.11.002
6. Caprì, S., Ignaccolo, M.: Genetic algorithms for solving the aircraft-sequencing problem: the introduction of departures into the dynamic model. J. Air Transp. Manag. **10**, 345 (2004). https://doi.org/10.1016/j.jairtraman.2004.05.004

7. Eurocontrol Experimental Centre: Point Merge Integration of Arrival Flows Enabling Extensive RNAV Application and CDA - Operational Services and Environment Definition, ATC Operating System, no. 2.0, pp. 1–89 (2010)
8. Eurocontrol Experimental Centre: Valdation Report - Real Time Simulation Dublin TMA2012, Implementation of a Point Merge System in Dublin, November 2010
9. Eurocontrol: Point Merge System Results of ENAV Prototyping Sessions (2008)
10. Hong, Y., Choi, B., Lee, S., Lee, K., Kim, Y.: Optimal and practical aircraft sequencing and scheduling for point merge system. IFAC-PapersOnLine **50**, 14644 (2017). https://doi.org/10.1016/j.ifacol.2017.08.1904
11. ICAO, Continuous Descent Operations (CDO) Manual (2010)
12. International Civil Aviation Organization, Doc 4444 - Air Traffic Management - Procedures for Air Navigation Services
13. Kwasiborska, A., Skorupski, J.: Analysis of the process of merging air traffic streams. Case study of TMA Warsaw. In: Communications in Computer and Information Science, vol. 897, pp. 320–334 (2018). https://doi.org/10.1007/978-3-319-97955-7_22
14. Kwasiborska, A.: Sequencing landing aircraft process to minimize schedule length. Transp. Res. Procedia. **28**, 111 (2017). https://doi.org/10.1016/j.trpro.2017.12.175
15. Lawler, E.L., Lenstra, J.K., Rinnooy Kan, A.H.G., Shmoys, D.B.: Sequencing and scheduling: algorithms and complexity. Handb. Oper. Res. Manag. Sci. **4**, 445 (1993). https://doi.org/10.1016/S0927-0507(05)80189-6
16. Lee, S., Hong, Y., Kim, Y.: Optimal Scheduling Algorithm in Point Merge System Including Holding Pattern Based on MILP (2017). https://doi.org/10.13009/EUCASS2017-137
17. Liang, M., et al.: Potential Operational Benefits of Multi-layer Point Merge System on Dense TMA Operation Hybrid arrival trajectory optimization ap-plied to Beijing Capital International Airport To cite this version: HAL Id : hal-01355018 Potential Operational Benefits of Multi-layer Point Merge System on Dense TMA Operation Hybrid Arrival Trajectory Optimization Applied to Beijing Capital International Airport (2016)
18. Lieder, A., Stolletz, R.: Scheduling aircraft take-offs and landings on interdependent and heterogeneous runways. Transp. Res. Part E Logist. Transp. Rev. **88** (2016). https://doi.org/10.1016/j.tre.2016.01.015
19. Mesgarpour, M.: Airport Runway Optimization, Ph.D. thesis Faculty Soc. Hum. Sci. Sch. Math. Univ. Southhampt. (2012)
20. Report, Real Time Simulation Oslo ASAP (Oslo Advanced Sectorization and Automation Project), no. January (2010).
21. SESAR Consortium, European ATM Master Plan - Edition 2, Roadmap Sustain. Air Traffic Manag. (2012). https://doi.org/10.2829/512525
22. Skorupski, J.: Airport traffic simulation using petri nets. In: Communications in Computer and Information Science (2013). https://doi.org/10.1007/978-3-642-41647-7_57
23. Skorupski, J., Florowski, A.: Method for evaluating the landing aircraft sequence under disturbed conditions with the use of Petri nets. Aeronaut. J. **120**, 819 (2016). https://doi.org/10.1017/aer.2016.32
24. Soomer, M.J., Franx, G.J.: Scheduling aircraft landings using airlines' preferences. Eur. J. Oper. Res. **190**, 798 (2008). https://doi.org/10.1016/j.ejor.2007.06.017
25. Tavakkoli-Moghaddam, R., Yaghoubi-Panah, M., Radmehr, F.: Scheduling the sequence of aircraft landings for a single runway using a fuzzy programming approach. J. Air Transp. Manag. **25**, 15 (2012). https://doi.org/10.1016/j.jairtraman.2012.03.004
26. Vakili, S., Liu, K., Zhao, Q.: Deterministic sequencing of exploration and exploitation for multi-armed bandit problems. IEEE J. Sel. Top. Signal Process. **7**, 759 (2013). https://doi.org/10.1109/JSTSP.2013.2263494

27. Vito, D., Vittorio, E.: Study of point merge technique for efficient continuous descent operations in TMA continuous descent operations in study of point merge technique for efficient study of point merge technique for efficient continuous descent operations. IFAC-PapersOnLine **51**(9), 193–199 (2018). https://doi.org/10.1016/j.ifacol.2018.07.032
28. Wilde, J.M. de: Implementing point merge system based arrival management at Amsterdam Airport Schipol (2018)

Assessing the Impact of a Potential Short-Haul Flights Ban on European Airports

Robert Szymczak[1,2](✉)

[1] Faculty of Transport, Warsaw University of Technology, Koszykowa 75, 00-662 Warsaw, Poland
robert.szymczak@wt.pw.edu.pl
[2] Arup Polska, Inflancka 4, 00-189 Warsaw, Poland
robert.szymczak@arup.com

Abstract. Increasing environmental awareness is putting a growing pressure on modern aviation. This is particularly reflected in calls for the introduction of a ban on short-haul flights that can be replaced by rail transport, which is currently considered in the European Union and some other Western European countries. The article aims at estimating the impact of the short-haul flights ban on air traffic at European airports. The analysis was performed for no-fly scenarios on routes where alternative rail travel takes no more than 3, 4, 5 or 6 h, respectively. The paper estimates the number of aircraft movements and air transport seats that would be affected by a potential ban on short-haul flights. Considering the impact on the environment, an estimation has been made to assess the extent to which the number of flight hours could be limited on all restricted routes. Performed analyses enabled the identification of flights which could be replaced by rail connections, broken down by countries and airports.

Keywords: Aviation market · Airports · Rail · Environment · Short-haul flights ban

1 Background

1.1 The Environmental Impact of Air Transport

Similar to all human activities, air transport produces carbon dioxide (CO_2) emissions as it uses fossil fuel to power aircraft. Aircraft are the fastest widely used mean of transport, which allows people and goods to reach nearly any place on earth within just a day. However with high speeds comes the significant contribution of aviation to air pollution and climate change. Air transport accounts for approximately 2% of global man-made carbon dioxide (CO_2) emissions [18] and 3% of EU's total greenhouse gas emissions [11]. If the global aviation sector was treated as a country, it would be in the top 10 CO_2 emitters.

In order to reduce the negative impact of aviation on the environment, the ICAO and the IATA have set a goal of carbon-neutral aviation growth since 2020 and a reduction in total net aviation CO_2 emissions of 50% by 2050, relative to 2005 levels [18].

© The Author(s), under exclusive license to Springer Nature Switzerland AG 2021
A. Kwasiborska et al. (Eds.): ATE 2020, LNITI, pp. 146–163, 2021.
https://doi.org/10.1007/978-3-030-70924-2_12

Flights over 1500 km without practical alternatives are responsible for around 80% of aviation CO_2 emissions [4]. Reducing the carbon footprint can be achieved mainly by replacing aircraft types with more efficient models, and optimizing ATC and airport procedures. Switching from older aircraft types like Airbus A340 or Boeing 767 to new such as Airbus A350 or Boeing 787 can cut down the emissions by 15–20% when flying on the same route [24].

The reduction of negative environmental impact of flights under 1500 km (which account for 20% of global air traffic) can be boosted by replacing some of the flights with rail or bus connections. Emissions measured in terms of CO_2 per passenger km of each mode of transport is highly dependent on the engine and fuel type, vehicle capacity and utilisation and efficiency. To provide a general comparison, Table 1 presents the average direct carbon dioxide emission per vehicle type.

Table 1. Average direct carbon dioxide (CO_2) emissions per vehicle type [7, 9, 15, 27]

Vehicle	Road vehicles			Train		Aircraft	
	Car 1 pax	Car 4 pax	Coach	average	Eurostar	average	E195-E2
Emission [gCO_2/pkm]	152	38	36	28	6	88	65

On average, train and long-distance coach services produce approx. 3 times less CO_2 emissions per passenger-kilometer than an average aircraft with approximately 80% load factor. The carbon footprint of electrically powered railways depends on the primary energy used to generate the electricity [12, 32]. Some of the rail operators such as Eurostar can go down to 14 times less CO_2 emissions than air travel, due to high train capacity and load factor, use of electricity provided by nuclear plants, better driving techniques and more efficient onboard lighting and heating [7, 31]. Similar to trains, some modern aircraft also have lower emissions than the average for airplanes. An example is the new Embraer E195-E2 with a maximum capacity of 146 passengers, which has only 2 times higher CO_2 emissions per passenger-kilometer than an average train or coach but still nearly 11 times more than the most efficient trains such as Eurostar [7, 9, 27]. The least CO_2 efficient vehicles are cars with a single passenger, producing almost 2 times more direct emissions per passenger than an aircraft and up to 25 times more than the Eurostar train.

1.2 The Idea of Short-Haul Flights Ban

The dynamic growth of aviation and high CO_2 emissions of air travel compared to rail became a trigger for discussion of limiting or completely banning short-haul flights. This idea has gained popularity especially in the European Union, where the railway network is well developed and many flight routes duplicate high speed train connections.

In 2018, pro-ecological organisations and activists such as Greta Thunberg started promoting the “flight shame” movement, named “flygskam,” in Swedish, “flugscham” in German, or “vliegschaamte” in Dutch. “Flight shame” is the feeling of guilt over the environmental effects of flying, avoiding flights and encouraging others to do the same [23]. Frans Timmermans, the European Social Democrats' candidate for the May 2019 EU elections, publicly stated that he would support a total ban on short-distance flights. Also his conservative counterpart Manfred Weber advocated for a reduction of such flights, although not supporting a complete ban [16].

According to a survey of the European Investment Bank (EIB), a majority of European citizens would support a ban on short-haul flights to fight climate change. Out of 28 088 respondents to the survey, 62% favored a ban and an even greater majority of 72% said they would support a carbon tax on flights [1]. The poll, conducted in September-October 2019, covered the then-28 European Union member states, including Great Britain.

European Commission President Ursula von der Leyen has said all industry sectors must contribute to the EU's target to reduce the bloc's net emissions to zero by 2050. Some airlines have started voluntarily offsetting emissions using carbon credits, although it is not clear whether the EU would accept these as a contribution to its 2050 goal [1]. In order to reduce greenhouse gas emissions, the KLM Royal Dutch Airlines replaced one of its five daily services between Brussels and Amsterdam Airport Schiphol with seat capacity aboard the Thalys high-speed train [19]. In the following years, the KLM plans to replace more flights with rail connections based on the cooperation with NS Dutch Railways and Thalys [19].

A potential short haul flights ban could be initially introduced by countries on a national level, followed by a ban for larger regions such as the European Union. Different levels of the ban are possible, such as:

- ban of all flights where a rail alternative under 2/2.5/3/4 etc. hours is available,
- same as above but only restricted to point to point passengers, flights for transfer passengers would still be allowed,
- complete ban of all domestic flights within a single country/within the EU,
- ban of flight routes where daytime or sleeper trains are available [29].

The recent COVID-19 pandemic has noticeably boosted the process of replacing flights with rail connections. One of the conditions of a 7 billion-euro state aid for Air France in May 2020 was to drastically reduce the number of domestic short haul flights, where a 2.5 h rail connection is available, by restricting their capacity to exclusively passengers transferring for another flight [30][34]. For example, only passengers connecting via Paris CDG or ORY from outside of the country will be allowed to fly from Paris to places like Lyon, Bordeaux, and Rennes, where fast TGV train operate [5, 36]. The French government also plans to eventually ban all other airline carriers operating domestic flights [20].

Poland plans to replace most of its domestic flights with rail connections by 2027, when the new Solidarity Transport Hub (STH) will be opened in Baranów near Warsaw [6]. The STH will be a "greenfield" multimodal transport hub consisting of an airport with an initial capacity of 45 MPPA (million passengers per annum) and a railway hub in a hub and spoke network, providing fast rail connections across the country thanks to over 1700 km of planned new railway lines. The project is currently in the planning phase with first construction works planned for 2023 and the airport opening in 2027 [6, 35].

1.3 Aim of the Article

Although the idea of banning short haul flights is becoming more and more popular in order to quickly reduce the environmental footprint of aviation, still no complex studies have been made to assess the impact of such a law on airports, airlines, passenger connectivity and the environment. In early 2000s, the idea of replacing flights with train connections was considered as a potential solution for reducing airport congestion. In 2003, the impact of replacing short haul flights with rail was analysed by Lopez-Pita A. and Anton F.R. for major European airports [22]. The conclusion was that due to the replacement of domestic flights and flights up to 400 km with high speed rail, up to 54 000 slots can become available at Madrid Airport, followed by around 40 000 slots at Paris CDG and 20 000 at Frankfurt Airport. Since then many scientific papers were published discussing the competition and cooperation between the EU air transport and rail, covering categories such as travel time and ticket price [33], passenger preferences [13], different airline and rail operator business models [2, 13], seats and flight frequencies [8], and environment and social welfare [33]. However none of these articles quantified the estimated effect of a potential short haul flights ban across Europe in terms of the number of flights and passengers affected, as well as the scale of air traffic movements and flying time reduction in relation to the overall traffic.

Implementation of any flights ban is likely to cause protests of the aviation industry and generate legislation problems. Such a case can be now observed in France where the Union of Autonomous Airlines (SCARA) protested against the partial ban of domestic flights proposed by the French government [3]. According to SCARA, there is no scientific data proving the benefits of such a ban and implementing it would infringe constitutional freedoms.

This paper quantifies the estimated effect of a potential short haul flights ban across Europe, focusing on the impact on airports, assuming that such a ban would be successfully implemented. The ultimate aim of this article is to provide data for further discussions on implementing, modifying or abandoning laws replacing flights with rail transport.

2 Methodology

2.1 Introduction

Europe is one of the regions where the process of replacing flights by rail is ongoing and where a potential domestic/short haul flights ban is discussed – especially in the European Union and neighboring countries such as Great Britain.

The article estimates the impact on airports of a potential short haul flights ban in the region of Europe, taking into consideration the following countries:

- 27 EU members: Austria (AU), Belgium (BE), Bulgaria (BG), Croatia (HR), Cyprus (CY), Czech Republic (CZ), Denmark (DK), Estonia (EE), Finland (FI), France (FR), Germany (DE), Greece (GR), Hungary (HU), Ireland (IE), Italy (IT), Latvia (LV), Lithuania (LT), Luxembourg (LU), Malta (MT), Netherlands (NL), Poland (PL), Portugal (PT), Romania (RO), Slovakia (SK), Slovenia (SI), Spain (ES), Sweden (SE).

- 3 non-EU Western European countries: Norway (NO), Switzerland (CH), Great Britain (with Northern Ireland) (GB).

The identification of flight routes affected by a potential ban was done for the calendar year 2019, and conducted in three steps, described in the following chapters:

- Preparation of a flights database (Sect. 2.2),
- Creation of a rail connections database (Sect. 2.3),
- Definition of flight banning rules (Sect. 2.4).

2.2 Flights Database

Historic flights data for the calendar year 2019 was obtained from the OAG Schedules Analyser [26]. OAG is a global air travel data provider which offers access to future and historical airline schedules, ticket bookings, connections data and more. The database is widely used by airports, airlines, aviation consultants, financial & investment companies, government agencies [25]. The exported historic passenger flights dataset contained the following variables: departure airport, departure city, departure country, arrival airport, arrival city, arrival country, sum of flights, sum of seats, great circle distance, sum of flying time – for all worldwide flights to/from the analysed European countries. OAG covers nearly all scheduled passenger flights, however based on author's experience, it lacks some of passenger charter flights. According to Eurocontrol, charter flights in 2018 had just a 3.9% market share of IFR flights, compared to 82.8% of other scheduled passengers flights (the remaining 13.3% were cargo, military and business aircraft flights) [10]. Based on this fact, it was assumed that the OAG database is sufficiently accurate for estimating the total passenger traffic, however it must be noted that some of the traffic (approx. 3–5%) is not covered.

2.3 Rail Connections Database

A database of railway connections was created using the Google Maps Distance Matrix API. Google Maps was used as a source of rail schedules from operators of all of the considered countries (excluding Cyprus and Malta, where no scheduled train services are offered) [14]. The API train connections requests were sent for all city pairs from the flights database (based on OAG city name and country of origin/destination), for routes not exceeding 1500 km. Train connections were searched to/from city centers, as they were considered as the ultimate destinations, unlike airports. The train connections data was collected for Tuesday 29th of October 2019, which was assumed to be an average business day in Europe, outside the holiday season. In order to find the shortest travel duration, 8 departure times were checked (the API returned data of the nearest departing train): 06:00, 08:00, 09:00, 12:00, 15:00, 18:00, 20:00 and 22:00 local time. Train interchanges were allowed in order to get more connection options. The final database contained railway connections for corresponding flight routes, with the min. Train travel duration and the travel duration for each of the 8 times of the day checked.

2.4 Flight Ban Assumptions

The identification of routes affected by the potential ban was conducted in Microsoft Excel and Power BI for 4 no-fly scenarios, depending on the criterion of the maximum train travel duration replacing the flight: 3 h, 4 h, 5 h and 6 h. Such limits would still allow one day business trips and are the most likely to be considered by governments. Both domestic and international flights were analysed using the flights database. Train travel duration was based on the minimum time from the database relevant for 2019 and did not consider future improvements. Train frequency was not used as a ban criterion as it was assumed that it could increase in case of a flights ban. E.g. the 3 h scenario assumed that all flights with a corresponding train connection under 3 h were banned and completely replaced by rail (required at least 1 train in the database with travel duration less than or equal to 3 h). Ban implementation aspects were not analysed in the article, assuming for study purposes that such a law could be successfully implemented. Figure 1 presents the railway network map and passenger flight routes in the analysed area in 2019.

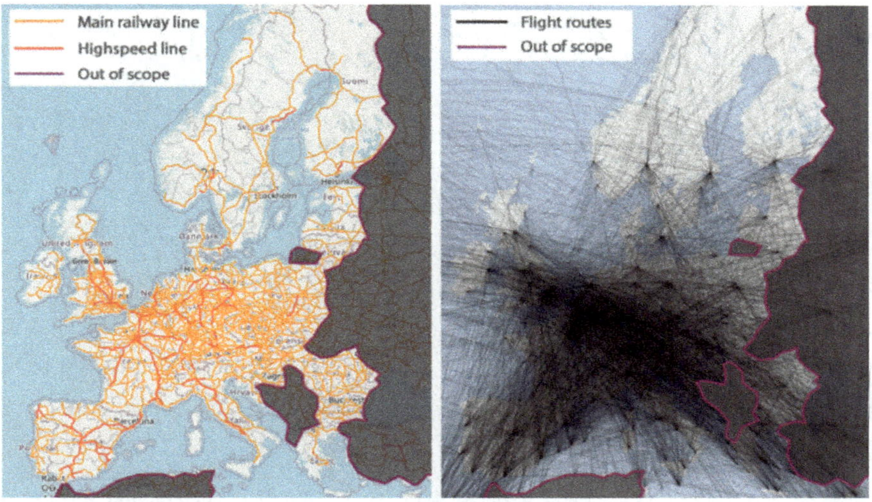

Fig. 1. Railway network map and passenger flight routes in the analysed area in 2019 (source: own work based on [26, 28])

3 Results

3.1 Identified Routes

Over 9 000 passenger air routes, which were operated from European airports in 2019 were analysed and compared with the rail connections database (Fig. 1) in order to identify those that could be replaced by rail. A map of identified routes for each no-fly

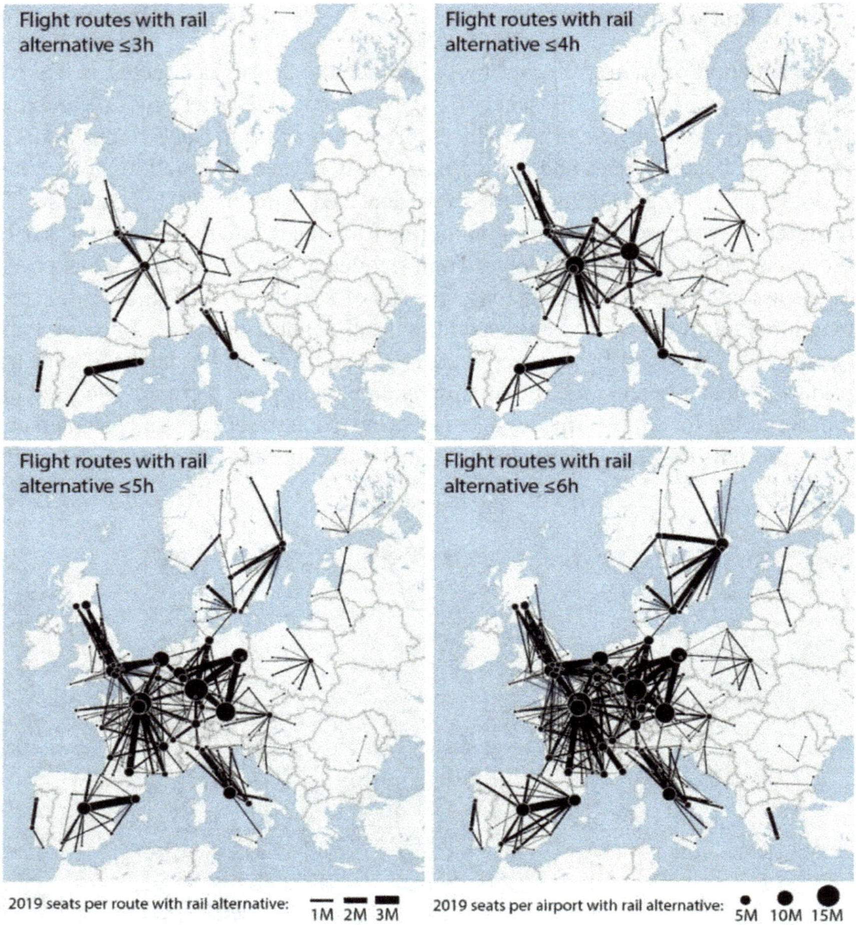

Fig. 2. Map of 2019 European flight routes with rail transport alternatives (source: own work)

Table 2. Summary of traffic affected by a potential short haul flights ban (source: own work)

Rail travel duration scenario	Airline traffic					Airport traffic		
	Routes	Flights (K)	Seats (M)	Flying time (h)	Distance (M km)	Airports	ATM (K)	Seats (M)
3 h	76	217.4	27.3	224 476	68.5	82	434.8	54.5
4 h	165	484.7	61.1	529 328	174.7	134	969.4	122.3
5 h	314	907.0	118.1	1 025 178	360.3	193	1 814.1	236.2
6 h	455	1 205.6	159.5	1 405 757	511.5	221	2 411.2	319.0
No rail <6 h	8 782	7 046.6	1 218.0	21 331 032	13 615.4	270	12 014.8	1 971.0
All traffic	9 237	8 252.2	1 377.5	22 736 789	14 126.9	491	14 426.0	2 290.0

scenario is presented below (Fig. 2). Results, definitions, abbreviations and airport code are described in Sect. 5.

Most of the identified air routes that could be affected by the short haul flights ban are located in Western Europe, especially between airports in France, Germany, Benelux countries, Switzerland, Austria and Great Britain. Other noticeable affected routes are within domestic markets in Italy, Spain, Sweden and Poland. When comparing Fig. 2 with Fig. 1, it can be noted that the majority of affected routes corresponds with a high-speed rail network (HSR). Several hub and spoke models are visible with centers in e.g. Paris, Rome, Frankfurt, Stockholm, Warsaw.

In the case of the 3 h ban, over 217 400 flights would not take place in Europe, which converts to 27 million airline seats (Table 2). For airports, the ban is equal to almost 435 000 less air traffic movements (ATM) and a reduction of 54 million airport seats at a total of 82 airports (departures and arrivals are counted separately). This would save around 224 000 flying hours. For the 4 h scenario, these values are approx. Doubled, for the 5 h scenario – 4 times higher, and for the 6 h scenario – multiplied by almost 6 (Table 2). In the 3 h scenario, 76 routes are affected, including the busiest route in Europe BCN-MAD (3.6 million seats, 19 000 flights), and routes between airports in the London and Paris region, as well as routes between Rome and Milan. In the 6 h scenario, 455 connections are banned. In addition to the 3 h scenario pairs, the busiest affected routes are MUC-TXL, FRA-TXL, ORY-TLS (Table 3).

Table 3. The busiest routes affected by a potential short haul flights ban (source: own work)

Route	Flights (K)	Airline seats (M)	Flying time (h)	Distance (M km)	Single flight duration (h)	Rail travel duration (h)
BCN<>MAD	19.0	3.6	25 391	9.1	1:25	2:45
MUC<>TXL	17.2	2.8	20 545	8.2	1:05	4:08
FRA<>TXL	16.5	3.1	19 543	7.1	1:10	4:09
ORY<>TLS	16.4	2.9	21 396	9.3	1:20	4:36
DUS<>MUC	14.7	2.4	17 250	7.1	1:10	4:43
NCE<>ORY	14.0	2.5	20 395	9.4	1:30	5:45
AMS<>LHR	13.4	2.1	17 919	4.9	1:20	4:42
ARN<>OSL	12.9	2.3	13 009	5.0	1:00	5:59

Looking at the whole air traffic covered by the study, the 3 h train travel duration scenario would cause a reduction of 1% of air routes to/from European airports, 3% of flights, 3% of airline seats and a 1% reduction of the total flying time (Fig. 3). This is because the affected routes are usually short distance, however operated with high frequency (often several flights a day). Out of all analysed European airports, 17% would be affected by the ban. The reduction of the negative impact of aviation on the environment could be represented by the 3% reduction of air traffic movements at airports (ATMs) and 1% reduction of flying time.

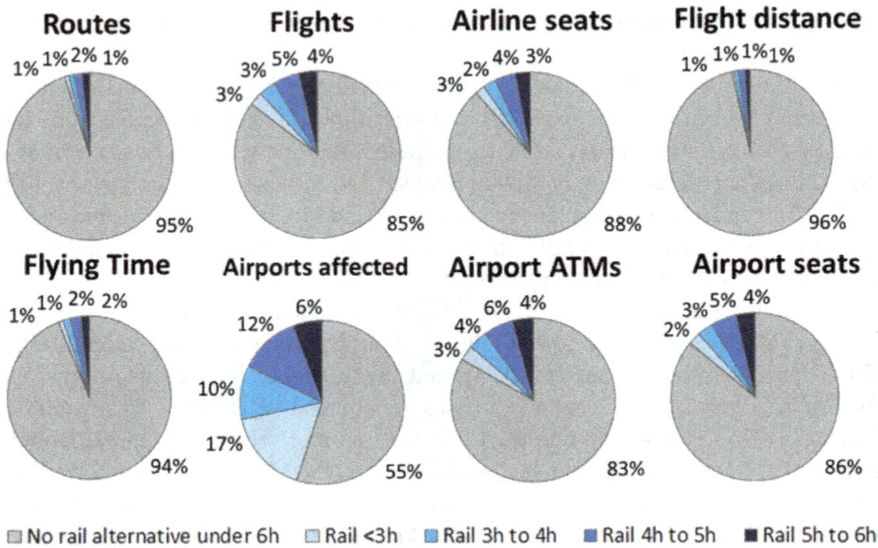

Fig. 3. Share of affected traffic by the short haul flights ban (source: own work)

As shown in Fig. 3, the air traffic reduction increases with the train travel duration criterion. The biggest gain is visible between the 4 h and 5 h scenario, when a number of busy domestic routes in Germany and France is banned.

In case of the 6 h rail travel duration scenario, a total of 5% routes would be banned, which means a 15% reduction of flights, a 12% reduction of airline seats, a 4% reduction of flight distance and a 6% reduction of flying time. Exactly 45% of European airports would be impacted by the ban, airport ATMs would be reduced by 17% and airport seats by 14%. The environmental impact could be noticeably reduced, taking into consideration the potential 6% reduction of flying time and 17% reduction of ATMs.

3.2 Ban Impact by Airport

The potential short haul flights ban would hit airports of all sizes across Europe. The biggest absolute traffic decrease would be observed at major hubs such as Paris Charles De Gaulle (CDG), Frankfurt (FRA) and Munich (MUC) (Fig. 4). The 3 h scenario would cause reductions up to 30 000 ATMs at CDG, FCO, MAD which equals to around 5 million annual seats at each of these airport (Table 4). The 4 h scenario noticeably affects CDG, FRA cutting in total around 77 000 ATMs and nearly 12 million seats at Paris CDG and Frankfurt. The 6 h scenario further increases reductions, going to over 115 000 ATMs and 17–19 million seats reduced at CDG and FRA.

The level of traffic reduction for each no-fly scenario highly depends on the airport and its city location in relation to high speed rail network. In case of Paris CDG, which is centrally located in the French railway system, the increase of banned air traffic is similar for each no-fly scenario. Contrary to CDG, at Berlin Tegel (TXL), there is almost no traffic reduction with the 3 h and 4 h no-fly scenarios. However, the 5 h train travel duration scenario would suddenly ban over 56 700 ATMs and 9.4 million seats at TXL

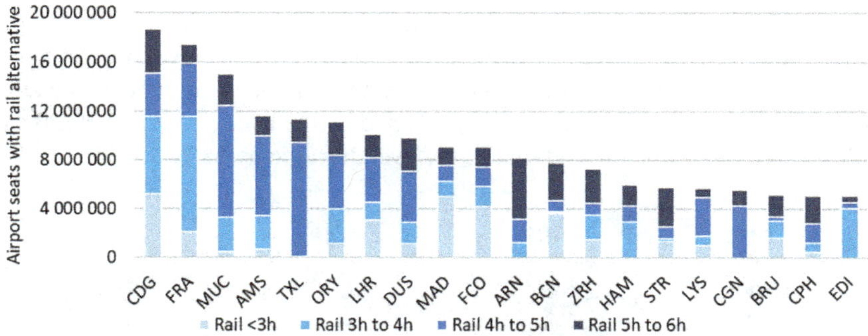

Fig. 4. Top 20 most affected airports by the potential short haul flights ban (source: own work)

(Table 4). The reason of this is that Berlin is located on the north-east end of the German railway network, and most of the rail connections to densely populated areas in the west and south of the country take between 4 h and 5 h. A similar case can be observed at Stockholm Arlanda (ARN), Cologne Bohn (CGN) and Edinburgh (EDI).

Table 4. Air traffic affected by a potential short haul flights ban at selected airports (source: own work)

Airport	ATM (K)					Seats (M)				
	Total	Max alternative rail travel				Total	Max alternative rail travel			
		3 h	4 h	5 h	6 h		3 h	4 h	5 h	6 h
AMS	476.1	7.2	25.0	71.0	83.9	82.1	0.7	3.5	10.0	11.6
ARN	216.3	0.0	8.2	21.8	52.1	33.6	0.0	1.3	3.2	8.2
BMA	44.6	0.0	6.6	25.7	27.9	3.6	0.0	0.6	2.1	2.3
CDG	466.5	31.2	76.8	101.8	126.1	90.2	5.3	11.6	15.1	18.7
EAS	4.7	0.0	0.0	4.7	4.7	0.4	0.0	0.0	0.4	0.4
FCO	306.2	31.8	41.2	51.3	61.2	56.0	4.2	5.9	7.5	9.1
FRA	483.9	16.0	77.7	103.0	117.1	90.1	2.1	11.6	15.9	17.4
LHR	478.0	19.7	30.5	54.3	65.3	100.3	3.1	4.6	8.2	10.2
MAD	398.3	32.7	45.1	55.7	66.6	72.6	5.0	6.3	7.6	9.1
MUC	394.8	5.7	24.7	84.5	104.5	63.4	0.5	3.3	12.5	15.0
TXL	187.9	0.0	0.4	56.7	71.3	30.8	0.0	0.1	9.4	11.4
WAW	170.9	16.0	23.9	24.7	27.4	21.2	1.4	2.1	2.2	2.4

The 3 h scenario ban would cause a reduction of traffic of up to 10% at airport hubs (FCO, MAD, WAW) (Fig. 5). However at the same time, a majority of airports would not face any reductions, or reductions would be negligible (1% at AMS, MUC). The 6 h ban could cause a reduction of up to 25% of traffic at large airports (CDG,

FRA), and various reductions at point-to-point airports. Approximately 20 airports in Europe would have all of their scheduled passenger flights removed by the ban (e.g. San Sebastián Airport EAS, based on the 2019 network). Many airports in Sweden with domestic flight networks would be highly affected by the ban, e.g. Stockholm Bromma (BMA), where the 6 h ban would reduce the overall traffic by 63%.

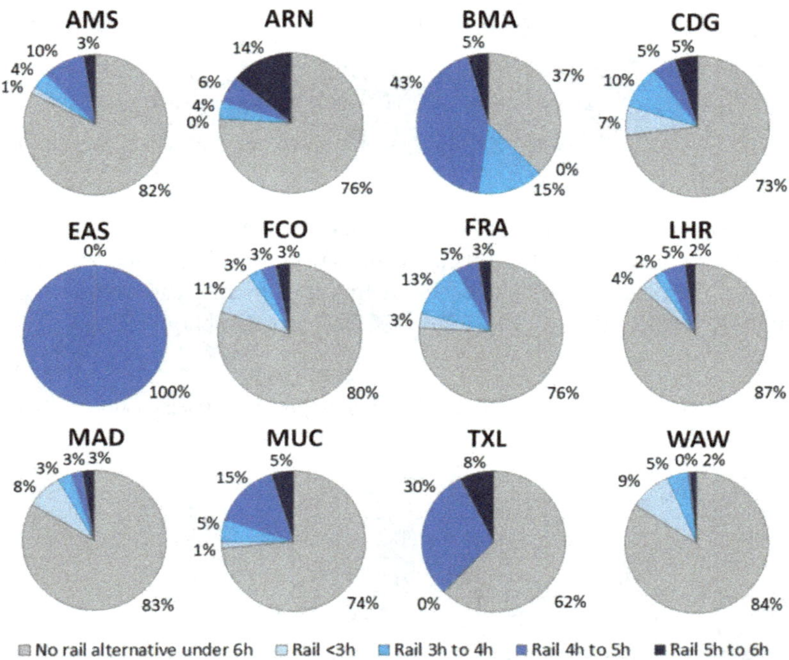

Fig. 5. Share of 2019 air traffic movements with rail alternative at selected European airports (source: own work)

Generally out of all 491 analysed airports, 55% would not be affected by the 6 h ban (Fig. 5). This ratio is higher for smaller airports with up to 1 million annual scheduled seats (68%) and much lower for large airports with over 10 million scheduled seats (15%). Large airports most often had a reduction of seats between 1–20%, followed by 21–41% reductions. Reductions of traffic at medium size airports were not higher than 61–80%, (e.g. BMA, MMX, MPL, AAL). Approximately 11.5% of small airports faced traffic reductions above 80% (36 airports, mainly in Sweden, Finland, Spain and France) (Fig. 5).

3.3 Ban Impact by Country

The potential short haul flights ban would affect a majority of European countries, hitting countries with well-developed high speed rail systems the most. In total, 8 European countries would observe a reduction of over 10 million annual seats at their airports. The biggest traffic drop would be noted in Germany and France, especially if the 6 h

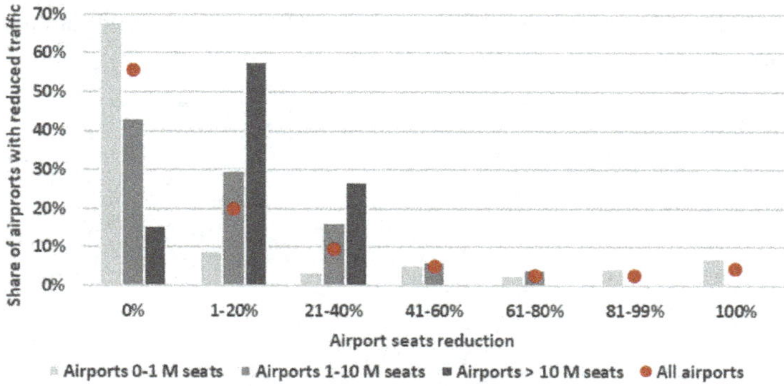

Fig. 6. Share of airports with reduced traffic in the 6 h ban scenario (source: own work)

train travel duration criterion would be implemented - approximately 65–80 million seats would be reduced at airports in each country (Fig. 7). Great Britain, which is the biggest aviation market in Europe, would be only the third most affected country, due to its limited high speed railway system and less central location in Europe (all international rail connections have to go through the Eurotunnel). It is worth highlighting that Sweden, with a relatively small aviation market, would see a significant reduction of traffic, being the 6th most affected country.

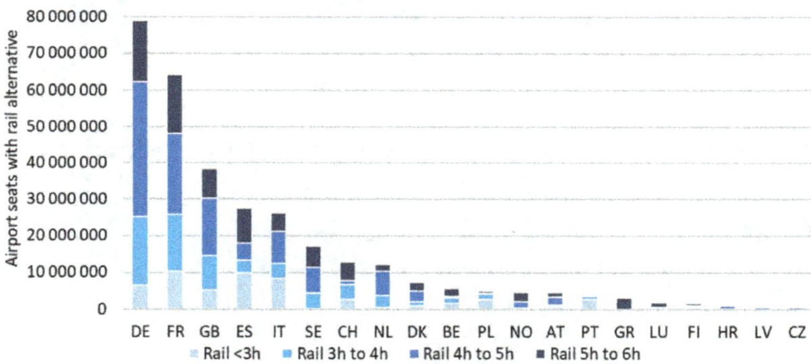

Fig. 7. Top 20 countries most affected by the potential short haul flights ban (source: own work)

In case of the 3 h ban scenario, France would face the biggest traffic reductions, followed by Spain and Italy (Table 5). Among these countries, traffic reductions would reach up to 80 000 ATMs and 10 million seats. Poland would be the 6th most affected country, as a number of its busiest domestic routes would be banned (e.g. WAW-GDN, WAW-KRK, WAW-WRO).

Looking at the share of air traffic affected by the 6 h ban scenario - in Luxembourg, Sweden and France, approximately 33% of ATMs would be replaced by rail (Fig. 8). Spain would be at the opposite end with a traffic reduction of only 10%. Flying time in case of the 6 h ban could be reduced by 22.5% in Luxembourg, 20% in Sweden, 17.5% in France, 13.5% in Germany and just 5–7% in Spain, Italy, Poland and Great Britain.

Table 5. Air traffic affected by a potential short haul flights ban in selected European countries (source: own work)

Country	ATM (K)					Airport Seats (M)				
	Total	Max alternative rail travel				Total	Max alternative rail travel			
		3 h	4 h	5 h	6 h		3 h	4 h	5 h	6 h
Belgium	257.9	11.5	23.7	29.4	42.5	42.9	1.6	3.1	3.7	5.6
Denmark	310.3	11.9	27.2	48.4	60.8	45.9	1.0	2.1	5.0	7.3
France	1 460.3	80.1	202.4	361.3	472.3	233.6	10.4	25.8	48.2	64.2
Germany	1 920.6	56.9	185.2	432.2	564.5	316.9	6.5	25.2	62.2	79.1
Great Britain	2 134.6	33.5	108.0	232.6	295.1	345.4	5.2	14.6	30.2	38.3
Italy	1 394.9	64.5	87.9	145.1	178.7	235.9	8.5	12.4	21.1	26.2
Luxembourg	54.2	4.0	4.0	7.7	19.1	6.1	0.3	0.3	0.6	1.9
Netherlands	533.9	7,2	28.1	74.2	91.8	92.1	0.7	3.8	10.3	12.3
Poland	367.2	32.1	47.7	49.4	55.0	52.0	2.8	4.2	4.3	5.0
Spain	1 880.8	65.5	96.2	128.5	188.2	310.5	10.0	13.3	17.9	27.5
Sweden	440.7	2.0	39.7	105.7	146.8	59.0	0.1	4.5	11.4	17.1
Switzerland	464.9	20.1	51.8	60.7	98.9	75.7	2.7	6.7	7.8	12.9

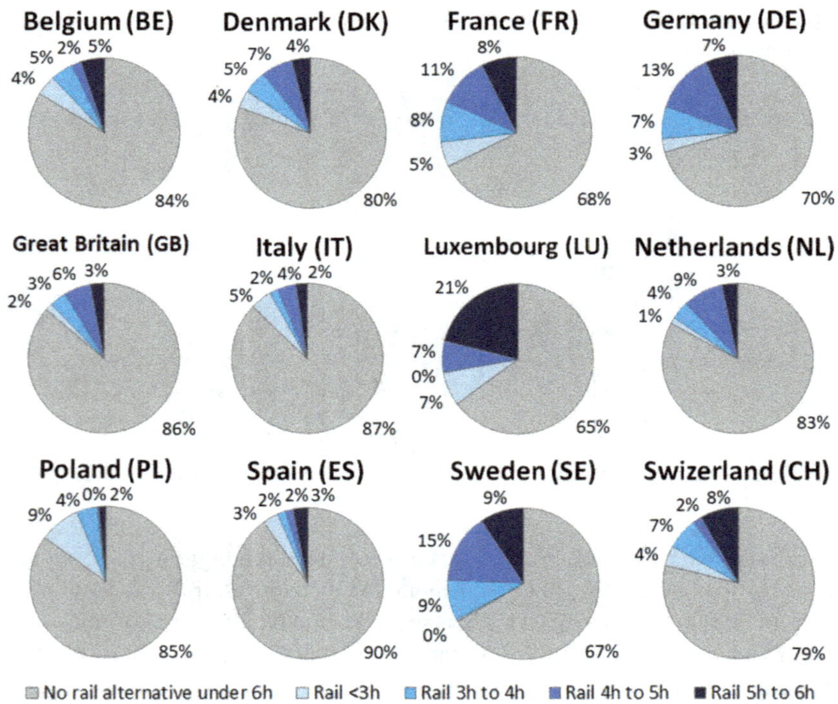

Fig. 8. Share of 2019 air traffic movements with rail alternative in selected European countries (source: own work)

The 3 h scenario would affect Poland the most, where 9% of all air traffic movements at airports would be banned. In a majority of other European countries traffic reductions would not exceed 5%. In Sweden, nearly no traffic would be affected by the 3 h scenario. Flying time reductions would stay as low as 4% in Poland, 2–3% in France and Italy, 1.5% in Spain and approx. 1% in Germany and Great Britain.

4 Conclusions

Reduction of the negative impact of aviation on the environment is one of the top priorities of the aviation sector and multiple international institutions. However, the proposed approach of banning selected short haul flights and replacing them with train connections, which are believed to be more ecologically friendly, is fairly controversial. One of the key parameters of such a ban is the maximum train travel duration, which would be used as the filter for air routes with alternative rail connections.

The implementation of a short haul flights ban in Europe with a relatively short maximum duration of train travel of 3 h would affect 17% of studied airports but hardly contribute to the reduction of the environmental impact of aviation – total flying time would be reduced by just 1%, air traffic movements reduced by 3%. The most affected airports would be the Paris Charles de Gaulle, Madrid Barajas and Rome Fiumicino (approx. 5 million seats banned). The biggest relative drop of air traffic would be seen in Poland (9% of ATMs), followed by Luxembourg and France.

The study showed that air traffic reduction increases with the train travel duration criterion. The results proved that a major difference is visible between the 4 h and 5 h scenario, when a number of busy domestic routes in Germany and France is banned.

The 6 h ban is likely to have a noticeable impact on the environment as the flying time would be reduced by 6% and ATMs would be reduced by 17%, at the cost of traffic reductions at 45% of airports. The most affected airports would be the Paris Charles de Gaulle, Frankfurt am Main and Munich, where up to 18 million seats would be banned. The biggest absolute passenger traffic reduction would be seen in Germany (almost 80 million seats), while relative reduction would be seen in in Luxembourg, Sweden and France (approx. 33%).

At the same time, it must be noted that probably some of the airport slots released due to the ban would be reused by airlines for launching new, longer routes - generating even more pollution than short flights. The absolute ban would cause serious difficulties for transfer passengers and potentially limit the connectivity of regional cities, which now often depend on air-links with hub airports. It would also likely force major airports to rearrange their facilities in a way to adjust processing areas for handling more passengers coming from the landside instead of the airside.

To sum up, the implementation of a short haul flights ban across all analysed countries would be a very drastic step with limited benefits, and based on the recent protests in France [3], it could be a long and difficult process. The 3 h and 4 h scenario would hardly contribute to the reduction of environmental pollution but at the same time affect up to a quarter of European airports. The analysis proved that the railway duration criterion set at 5 h or 6 h in theory could noticeably contribute to the reduction of environmental pollution, however only if new longer flights would not replace those banned (which is unlikely). Instead of completely banning flights, countries which already have a well developed railway market competitive with domestic flights (e.g. France, Sweden, Germany) could focus on promoting rail, implementing taxes on flight tickets competing with rail and offsetting aircraft emissions using carbon credits. Potential flight restrictions should first focus on origin-destination passengers to avoid limiting connectivity of regional airports benefiting from transferring flights to hub airports.

The article could be complemented by a detailed environmental impact study based on aircraft and train pollution models as well as financial analyses, which would estimate the benefits of the ban more accurately. Railway capacity studies are also needed to confirm that the rail system is capable of efficiently taking over the banned air traffic.

5 Definitions

5.1 General

- **Airline seats:** available seat capacity on departure of a certain flight.
- **Airport ATMs**: airport air traffic movements - sum of departures and arrivals at a certain airport (For intra-European routes 1 flight generates 1 ATM at each of two European airports. In case of inter-continental routes - one flight contributes to 1 ATM at European airports).
- **Airport seats:** sum of seat capacity on departing and arriving aircraft at a certain airport.
- **ATM:** air traffic movements.
- **EU:** the European Union.
- **Flight distance:** Great Circle Distance in kilometers. The summed distance of individual legs (stopping) flights.
- **Flight:** a single passenger air transport movement from Airport A to Airport B.
- **Flying time:** flight duration based on block times.
- **HSR:** High-speed rail network.
- **K:** thousand.
- **M:** million.
- **Rail travel duration:** the minimum train journey time on a certain route (city center to city center instead of airport to airport).
- **Route:** a pair of airports between which passenger flights were operated.

5.2 Airport Codes

See Table 6.

Table 6. Airport codes (Source: [26])

AAL	Aalborg, Denmark	KRK	Krakow John Paul II, Poland
AMS	Amsterdam Schiphol, the Netherlands	LHR	London Heathrow, Great Britain
ARN	Stockholm Arlanda, Sweden	MAD	Madrid Barajas, Spain
BCN	Barcelona El Prat, Spain	MMX	Malmo, Sweden
BMA	Stockholm Bromma, Sweden	MPL	Montpellier, France
CDG	Paris Charles de Gaulle, France	MUC	Munich, Germany
CGN	Cologne Bonn, Germany	NCE	Nice Cote d'Azur, France
DUS	Dusseldorf, Germany	ORY	Paris Orly, France
EAS	San Sebastian, Spain	OSL	Oslo Gardermoen, Norway
EDI	Edinburgh, Great Britain	TLS	Touluse-Blagnac, France
FCO	Rome Fiumicino, Italy	TXL	Berlin Tegel, Germany
FRA	Frankfurt Main, Germany	WAW	Warsaw Chopin, Poland
GDN	Gdansk Lech Walesa, Poland	WRO	Wroclaw Copernicus, Poland

5.3 Country Codes

See Table 7.

Table 7. Country codes (Source: [26])

AU	Austria	HU	Hungary
BE	Belgium	IE	Ireland
BG	Bulgaria	IT	Italy
CH	Switzerland	LT	Lithuania
CY	Cyprus	LU	Luxembourg
CZ	Czech Republic	LV	Latvia
DE	Germany	MT	Malta
DK	Denmark	NL	Netherlands
EE	Estonia	NO	Norway
ES	Spain	PL	Poland
FI	Finland	PT	Portugal
FR	France	RO	Romania
GB	Great Britain (with Northern Ireland)	SE	Sweden
GR	Greece	SI	Slovenia
HR	Croatia	SK	Slovakia

References

1. Abnett, K.: Ban short-haul flights for climate? In EU poll 62% say yes. Reuters (2020). https://www.reuters.com/article/us-climate-change-eu-flights/ban-short-haul-flights-for-climate-in-eu-poll-62-say-yes-idUSKBN20X2RA. Accessed 17 Mar 2020
2. Adler, N., Nash, C., Pels, E.: High-Speed Rail & Air Transport Competition, Tinbergen Institute Discussion Paper, TI 2008-103/3 (2008)
3. Air Journal France: Vols courts interdits en France : les acteurs de l'aérien protestant. https://www.air-journal.fr/2020-06-24-vols-courts-interdits-en-france-les-acteurs-de-laerien-protestent-5221038.html. Accessed 24 June 2020
4. Air Transport Action Group: Aviation and climate change (2020). https://aviationbenefits.org/media/166900/fact-sheet_2_aviation-and-climate-change.pdf. Accessed 28 Mar 2020
5. BFM, Air France sommée de ne plus vendre de billets entre Paris et une ville française située à moins de 2h30 en TGV. https://www.bfmtv.com/economie/air-france-sommee-de-ne-plus-vendre-de-billets-entre-paris-et-une-ville-francaise-situee-a-moins-de-2h30-en-tgv-1904372.html. Accessed 10 May 2020
6. Central Transportation Hub, New international airport for Poland, https://cpk.pl/en/about/airport-component. Accessed 10 May 2020
7. Department for Business, Energy & Industrial Strategy: 2019 Government greenhouse gas conversion factors for company reporting. https://assets.publishing.service.gov.uk/government/uploads/system/uploads/attachment_data/file/829336/2019_Green-house-gas-reporting-methodology.pdf. Accessed 28 Mar 2020
8. Dobruszkes, F., Dehon, C., Givoni, M.: Does European high-speed rail affect the current level of air services? An EU-wide analysis. Transp. Res. Part A **69**, 461–475 (2014)
9. Embraer Commercial Aviation: Feeling Flygskam? It's a Flying Shame. https://www.embraercommercialaviation.com/feeling-flygskam-its-a-flying-shame. Accessed 01 Feb 2020
10. Eurocontrol, EUROCONTROL Seven-Year Forecast February 2019. https://www.atceuc.org/uploads/docs/eurocontrol-7-year-forecast-february-2019-main-report.pdf. Accessed 18 Mar 2020
11. European Commission: Reducing emissions from aviation. https://ec.europa.eu/clima/policies/transport/aviation_en. Accessed 15 Mar 2020
12. European Environment Agency: Renewable energy in Europe – 2017 Update, EEA Report No 23/2017, Luxembourg (2017). https://op.europa.eu/en/publication-detail/-/publication/fa760a71-41eb-11e8-b5fe-01aa75ed71a1/language-en. Accessed 17 Mar 2020
13. Finger, M., Bert, N., Kupfer, D.: High-speed rail vs. low-cost air: competing or complementary modes?. European Transport Regulation Observer, Publications Office of the EU (2016)
14. Google Maps Transit, Cities Covered. https://maps.google.com/landing/transit/cities/index.html. Accessed 05 May 2020
15. Graver, B., Zhang, K., Rutherford, D.: CO_2 emissions from commercial aviation, The International Council on Clean Transport (2018)
16. Grebenjak, M.: Caps or Bans on short-haul/domestic flights. In: Kretzschmar, A. (ed.) International Conference on: Degrowth of Aviation, pp. 24–27. Stay Grounded Network - Kollektiv Periskop, Vienna (2019)
17. IATA, Airport codes. https://www.iata.org/en/publications/directories/code-search/?airport.search=krk. Accessed 20 July 2020
18. IATA, Fact Sheet: Climate Change & CORSIA (2018). https://www.iata.org/contentassets/c4f9f0450212472b96dac114a06cc4fa/fact-sheet-climate-change.pdf. Accessed 26 Feb 2020
19. KLM, KLM, Thalys and NS Dutch Railways have joined forces to replace flights between Brussels and Amsterdam Airport Schiphol. https://news.klm.com/klm-thalys-and-ns-dutch-railways-have-joined-forces-to-replace-flights-between-brussels-and-amsterdam-airport-schiphol/. Accessed 19 Feb 2020

20. Ledsom, A.: France To Permanently Ban Many Short-Haul Flights, Forbes. https://www.forbes.com/sites/alexledsom/2020/06/23/france-to-permanently-ban-many-short-haul-flights/. Accessed 23 June 2020
21. LNCS Homepage. https://www.springer.com/lncs. Accessed 21 Nov 2016
22. Lopez-Pita, A., Anton, F.A.: The effects of high-speed rail on the reduction of air traffic congestion. J. Public Transp. **6**(1) (2003)
23. Mkono, M.: Eco-anxiety and the flight shaming movement: implications for tourism, Journal of Tourism Futures (2020). https://www.emerald.com/insight/content/doi/10.1108/JTF-10-2019-0093/full/pdf?title=eco-anxiety-and-the-flight-shaming-movement-implications-for-tourism. Accessed 27 Mar 2020
24. National Academy of Sciences, Academy of Engineering, National Research Council, National: Real Prospects for Energy Efficiency in the United States, The National Academies Press, pp. 155–156, Washington D.C. (2010)
25. OAG, Schedules Analyser Sheet. https://www.oag.com/hubfs/Analyser/new_schedules_analyser/Schedules%20Analyser.pdf. Accessed 13 Feb 2020
26. OAG, Schedules Analyser Database. https://analytics.oag.com/analyser-client/home. Accessed 15 May 2020
27. Office of Rail and Road: Rail emissions 2018–19 Annual Statistical Release. https://dataportal.orr.gov.uk/media/1550/rail-emissions-2018-19.pdf. Accessed 28 Mar 2020
28. OpenRailwayMap. https://www.openrailwaymap.org/. Accessed 30 Apr 2020
29. Papa, E.: Could sleeper trains replace international air travel? The Conversation, London (2020)
30. Reuters, Air France must cut emissions, domestic flights for aid: minister. https://www.reuters.com/article/us-health-coronavirus-france-economy/air-france-must-cut-emissions-domestic-flights-for-aid-minister-idUSKBN22B2EL. Accessed 10 May 2020
31. Reynolds, L.: The Business Leader's Guide to the Low-carbon Economy, pp. 134–135. Routledge, New York (2012)
32. Servaes, T.S., Krautscheid, T., Schober, A.: A level playing field for comparing air and rail travel times. Open Transp. J. **13**(1), 48–56 (2019)
33. Sun, X., Zhang, Y., Wandelt, S.: Air transport versus high-speed rail: an overview and research agenda. J. Adv. Transp. (2017)
34. The Blue Swan Daily, Air France must reduce capacity on routes with rail alternatives: Minister, Air France must reduce capacity on routes with rail alternatives: Minister. https://blueswandaily.com/air-france-must-reduce-capacity-on-routes-with-rail-alternatives-minister/. Accessed 10 May 2020
35. The Council of Ministers of the Republic of Poland, Resolution no 173/2017 of 7 November 2017 on the adoption of the Investment Preparation and Implementation Concept: Solidarity Airport – Central Transport Hub for the Republic of Poland (2017)
36. View from the wing, Air France No Longer Permitted To Sell Domestic Trips That Could Be Taken By Train. https://viewfromthewing.com/air-france-no-longer-permitted-to-sell-domestic-trips-that-could-be-taken-by-train/. Accessed 10 May 2020

Author Index

© The Editor(s) (if applicable) and The Author(s), under exclusive license
to Springer Nature Switzerland AG 2021

A. Kwasiborska et al. (Eds.): ATE 2020, LNITI, p. 165, 2021.
https://doi.org/10.1007/978-3-030-70924-2

CPSIA information can be obtained
at www.ICGtesting.com
Printed in the USA
LVHW080818120321
680993LV00038B/81

9 783030 709235